EDIBLE
PLANTS
AND
ANIMALS

Other books by A.D. Livingston

Outdoor Life's Complete Game & Fish Cookbook
Good Vittles: One Man's Meat, a Few Vegetables, and a Drink or Two
Cast-Iron Cooking
Grilling, Smoking, and Barbecuing
Venison Cookbook

EDIBLE
PLANTS
AND
ANIMALS

Unusual Foods
from Aardvark to Zamia

A.D. Livingston
and
Helen Livingston, Ed.D.

 Facts On File

Edible Plants and Animals

Copyright © 1993 A. D. Livingston and Helen Livingston, Ed.D.

Facts On File, Inc.
460 Park Avenue South
New York NY 10016
USA

Library of Congress Cataloging-in-Publication Data

Livingston, A. D., 1932-
    Edible plants and animals : unusual foods from aardvark to zamia /
A.D. Livingston and Helen Livingston.
        p.   cm.
    Includes bibliographical references and index.
    ISBN 0-8160-2744-7 (hardcover)
    ISBN 0-8160-3051-0 (paperback)
    1. Animal food—Encyclopedias. 2. Plants, Edible—Encyclopedias.
I. Livingston, Helen. II. Title
TX743.L58   1993
641′.03—dc20   92-41501

A British CIP catalogue record for this book is available from the British Library.

Facts On File books are available at special discounts when purchased in bulk quantities for businesses, associations, institutions or sales promotions. Please contact our Special Sales Department in New York at 212/683-2244 or 800/322-8755.

Jacket design by Catherine Rincon Hyman
Composition by Grace M. Ferrara/Facts On File, Inc.
Manufactured by the Maple-Vail Book Manufacturing Group
Printed in the United States of America

10 9 8 7 6 5 4 3 2 1

This book is printed on acid-free paper.

CONTENTS

ACKNOWLEDGMENTS

The authors would like to thank the respective individuals, publishing houses, and magazines for their permission to use quoted material from their publications, as follows: *Magic and Medicine of Plants*, copyright 1986, the Reader's Digest Association, Inc.; *Cy Littlebee's Guide to Cooking Fish and Game*, copyright 1981, and *Wild Edibles of Missouri*, copyright 1979, by the Conservation Commission of the State of Missouri; Carleton S. Coon's *The Hunting Peoples*, published by Lyons & Burford, Inc.; Elizabeth Schneider's *Uncommon Fruits and Vegetables*, published by HarperCollins Publishers; *The South Carolina Wildlife Cookbook*, The University of South Carolina Press; *A Cruising Guide to the Tennessee River, Tenn-Tom Waterway, and Lower Tombigbee River*, by Mariam, Thomas W., and W.J. Rumsey, courtesy International Marine/McGraw-Hill, Camden, Maine; material reprinted with the permission of Atheneum Publishers, an imprint of Macmillan Publishing Company, from *Iron Pots and Wooden Spoons* by Jessica B. Harris, copyright 1989 by Jessica Harris; Reay Tannahill's *Food in History*, published by Crown Publishers Inc.; material from *The Atlantic Monthly* magazine, April 1992; material from several issues of *Outdoor Highlights*, published by the Illinois Department of Conservation; the Macmillan Publishing Company for material from *Insect Fact and Folklore* by Lucy W. Clausen; Dover Publications for material from *The Explorations of Captain James Cook in the Pacific*, edited by A. Grenfell Price; Andromeda Publishing for material from *The Encyclopedia of Mammals* and *The Encyclopedia of Reptiles*; and Charles Scribners and Sons, an imprint of Macmillan Publishing Company, for material from *Cross Creek Cookery* by Marjorie Kinnan Rawlings. The authors also acknowledge material from *The Travels of Marco Polo* published in the United States by Facts On File, Inc. and material from *The Histories of Herodotus* published by George Macy Co.

INTRODUCTION

Modern man usually eats a combination of meat and vegetables, just as he has done throughout his history. Although some early peoples were primarily hunters, they made good use of available berries, greens, and nuts. By the same token, others were primarily seed-gatherers or foragers, but they made good use of animals whenever the opportunity arose. For the most part, primitive man ate a wide range of foods, both animal and vegetable. As soon as the seed-gatherers started cultivating grain and the hunters started herding animals, however, man's range of foods began to narrow and his knowledge of the old ways began to fade. He began to specialize.

Machines for cultivating and harvesting continued the trend, making it very easy for man to obtain a few vegetables and grains that were ideal for mass production. Similar advances in animal husbandry, slaughter, and transportation, along with refrigeration, did the same thing with several animal foods. These days, the trend is toward meat that is packaged, wrapped, and displayed, and advertised as T-bone steak or hamburger without reference to the cow.

Whether or not man has become too narrow-minded in his eating habits—or perhaps too queasy of stomach for his own good—remains to be seen. Anyone who is abhorred by the idea of eating a monkey should know that these animals are still the favorite meat of human beings in some parts of the world. Further, they should very carefully consider the fact that some people who relish the taste of monkey may be shocked and even nauseated by the very idea of drinking cow's milk and might well ask rather pointed questions about what, exactly, is inside the American hotdog.

Although it covers some 500 rather unusual plants and animals that can be eaten, this work is not likely to change modern man's eating habits drastically. Only famine can do that. But the authors hope that the book will in some small way help preserve mankind's knowledge of edible foods, and that it will help broaden our cultural tolerance of each other's meat and potatoes. If any reader with adventurous taste chooses to experiment with the nutty crunch of the chufa in his breakfast cereal, or to cook Termites à la Bantu on the griddle, so much the better.

Part One

Edible Animals

Almost all animals—even highly poisonous snakes—are edible, or at least can be safely eaten by man. But there are exceptions, and some creatures, such as the puffer fish, can contain deadly parts while other parts might well be perfectly safe and even tasty.

Most of the alphabetically arranged entries that follow are the common names of various animals and, sometimes, of animal matter such as eggs or blood. If applicable, a scientific name immediately follows the headword or is somewhere in the text. For a case in which a particular animal has several common names, only one is given as a headword. Other names may be set forth in the text or are listed in the index. If in doubt, the reader should *always* consult the index. There can be a great deal of confusion, or misunderstanding, about the names of animals. The word "bass," for example, is used to denote a hundred different kinds of fish. For this reason, positive reference is best achieved by using the scientific names, which are also listed in the index.

A

aardvarks (*Orycteropus afer*) These long-snouted animals with a stout, piglike body live in the southern two-thirds of Africa. At home on the plains as well as in the forest, they feed for the most part on ants and termites, which they catch with the aid of a foot-long sticky tongue. They feed at night and rest in large undergrown burrows during the day.

The aardvark is highly prized for its flesh, which tastes like pork. In fact, its name is Dutch for "earth pig." It weighs up to 140 pounds, which makes it a worthwhile target for meat hunters. Also, the aardvark's teeth are considered to be good luck charms by some peoples in Zaire, and they are worn on necklaces.

aardwolves (*Proteles cristatus*) A member of the hyena family, the aardwolf lives in the drier parts of eastern and southern Africa. From head to tail, the aardwolf measures about 30 inches and resembles a long-nosed, striped wolf or fox. In fact, the name aardwolf means "earth wolf" in Afrikaans. The animals den in holes in the ground, which they either dig for themselves or borrow from porcupines and hares. An aardwolf may have 9 or 10 dens within its territory. The meat of the aardwolf is very tasty, and the animal is often used for food in Africa. Its striped pelt is also of local value.

Basically, the aardwolf is a shy creature, usually a solitary forager. During hard times, it eats various insects and even small birds, but the harvest termite makes up most of its diet. Apparently, it locates termites by an acute sense of sound. When termites are found, the aardwolf licks them up in short order with its long tongue, which is coated with a sticky substance. A single aardwolf can consume up to 20,000 termites in one night. Because harvest termites cause damage to livestock pastures, the aardwolf is important for pest control within its range.

abalone The tough foot of this large sea snail, which has a flattened shell, is highly regarded as table fare in various parts of the world. In England and France it is called *ormier*. In New Zealand, *paua*. In Australia, "ear shells." In South Africa, *perlemeon*. Once plentiful off the coast of California, abalone were at one time an important commercial seafood, but these days the commercial take is highly regulated. Abalone can no longer be canned in California, nor can they be exported from the state. Some local skin divers catch them in small numbers for private use.

The eight species of abalone that grow in California waters include the red (*Haliotis rufescens*), pink (*Haliotis corrugatau*), and green (*Haliotis fulgens*). A number of species grow around the world, and some are very popular in coastal China and in Japan, where girls dive after them.

The abalone gets mixed reviews from various quarters, being considered a delicacy by some people and as mere survival fare by others. A good deal depends on how it is prepared. The shell grows up to 10 inches in diameter, and the foot, when removed, looks like a steak. But it is very tough and requires much preparation before it can be eaten with ease. Usually, the abalone are tenderized only after much pounding, in a manner similar to how country cooks pound a beef-steak of inferior cut. After being tenderized, the abalone meat must be cooked quickly so that it won't again become rubbery and rather tasteless.

Not many years ago, the Maoris of New Zealand had a clever way of tenderizing *paua*. They buried it in the ground for several weeks—until it was soft—and then sliced it like cheese.

African pouched rats These edible rats comprise five species, all of which live in the southern half of Africa. One species, *Cricetomys gambianus*, grows up to 16 inches long (not counting the tail). These rats have a storage pouch that opens on the inside of each cheek. The pouches are used to carry food to the underground living quarters and storage chambers. The pouches can also be puffed up with air to make the creature look more forbidding. These rats feed on insects, fruit, seeds, and so forth—even snails.

The African pouched rat is quite plentiful south of the Sahara and they are sometimes hunted for food.

(See also RATS.)

Alaska blackfish (*Dallia pectoralis*) This small fish, a member of the mudminnow family, lives in Alaska and Siberia. Like other mudminnows, it is capable of living in poorly oxygenated waters. It survives well in cold weather, but has been incorrectly reported to be able to survive being frozen solid.

In any case, the Alaska blackfish grows to only 8 inches in length and has a slim body. In spite of its small size, however, it is sometimes very plentiful and is an important source of food for some people in northern Alaska and Siberia.

(See also FISH.)

albatross In the days of oceangoing sailing ships, the albatross was a special bird to sailors. Some people believed that to kill an albatross brought bad luck. The albatross was also associated with the wind; the birds were believed to be the spirits of seamen who had been washed overboard during storms. The idea of bad luck and a windless sea following the killing of an albatross was behind Samuel Taylor Coleridge's *The Rime of the Ancient Mariner*, in which the wind ceased to blow after the mariner shot the albatross with his crossbow. But most sailors had a rather irreverent attitude toward the birds, and called them "gooneys," the Old English word that has now been shortened to goon. Japanese seamen also called them *bakadori*, or fool-birds. The Portuguese navigators used the word *alcatraz* to mean a pelican or other large sea birds; in English this became albatross.

By whatever name, 13 species of the albatross soar over the southern seas from the Tropic of Capricorn to Antarctica, and in the Pacific north to the Bering Sea. They don't often frequent the North Atlantic or the frozen part of the Arctic Ocean.

Albatross have long, narrow wings as compared to the length of their body. The largest is the wandering albatross, *Diomedea exulans*, which has a length, from beak to toe, of 4 feet and a wingspan of over 11 feet. It is believed that this particular species can travel all the way around the world in less than 80 days.

Most albatross are famous for riding the trade winds and are also noted for following ships, often feeding in a ship's wake or on its garbage. Because of this habit, they are easy to catch with a baited hook trolled on the water's surface. They were sometimes caught for fresh meat, especially in the days before refrigeration. Although the birds are reported to have a strong flavor by modern standard, most any sort of fresh meat was probably a relief from the salt meat and hardtack that was eaten on sailing ships far from port. Sometimes the ships visited the albatross's nesting grounds, usually on isolated islands, and took both the birds and eggs for food.

But it was the feather hunters who, in the early 1900s, put a dent in the albatross population. In addition to the wings, the body feathers were used as "down" for stuffing pillows and mattresses. Reports of the albatross slaughter on Laysan Island, west of Hawaii, were so disturbing that Theodore Roosevelt made the island a wildlife reservation in 1909. During World War II, however, starving Japanese soldiers wiped out the rookery on Wake Island. Other rookeries have since been threatened by military and commercial aircraft.

(See also BIRDS.)

alligators and crocodiles Of the 22 species of alligators, crocodiles, caymans, and gharials that once thrived in the tropical and subtropical areas of the world, 19 species are now either threatened or endangered. Most of the problems have been caused by man, partly because of reduction of suitable habitat and partly because of a great demand for the hides of these reptiles, which are used in luxury leather products.

Surprisingly, the American alligator (*Alligator mississippiensis*) is doing very well in modern times. Not many years

ago, there was cause for concern for its future. But proper management techniques and rigid game protection laws helped the species rebound, and now they are present in such numbers that legal hunting is permitted in some states. Also, the large impoundments that have been built in the southeast and in Texas have increased the alligator's habitat, as have the warm water discharges from some nuclear electric plants along the larger rivers. Alligators are also raised on farms for hides and meat.

Alligator is very good table fare (if it is properly prepared) and is now available commercially in some fish or meat markets. It is also served in restaurants, and many alligator tail steaks are grilled at tailgate parties prior to the annual football game between the Georgia Bulldogs and the Florida Gators. At least from a culinary standpoint, the Georgia fans seem to have the best of this event. In any case, alligator tail steaks are traditional fare in parts of central Florida. Actually, the whole alligator is just as good as the tail, and maybe better for some purposes. The whitish meat usually has a mild flavor, not unlike frogs' legs, and is similar in texture to veal. The young alligators, less than five feet long, are usually considered to be better, or at least more tender, than the old ones.

In addition to the seven species of alligators, all 14 species of crocodiles are sometimes eaten, although some are now protected. In India, for instance, the protected mugger crocodile (*Crocodylus palustris*) is hunted illegally for meat and eggs.

amphioxi Also called arrowworms and lancelets, these strange sea creatures are found in clean sand under shallow water. They are small fellows, about 2 inches long, that look like eels or lampreys. The amphioxus usually stays buried unseen in the sand, but it is widely scattered about the tropics and temperate seas. Actually, there are a number of species in two genera, *Branchiostoma* and *Asymmetron*.

All of the species are not only edible but are (according to the *U.S. Armed Forces Survival Manual*) excellent either fresh or dried. Believe it or not, a commercial amphioxus fishery exists near Amoy, China, where the creature is especially plentiful.

anteaters Four species of these long-snouted animals can be found mostly in Central and South America, with a few in southern Mexico and Trinidad. As their common name implies, they all live mostly on ants and termites. Anteaters are sometimes used as food, though not often. Their hide is used in local leather goods, but their greatest commercial value seems to be as odd-looking trophies or specimens. The largest of the group—the giant anteater (*Myrmecophaga tridactyla*)—weighs up to 86 pounds.

Similar long-snouted ant-eating animals live in other parts of the world; some of these are highly rated as food.

(See also AARDVARKS; AARDWOLVES; ECHIDNAS; PANGOLINS.)

antelope The name "antelope" applies to many members of the family Bovidae, and generally includes everything that is not a cow, deer, bison, buffalo, sheep, or goat. In short, the zoological classification of the antelope is really a quagmire. In general terms, however, most of the antelope are distinguished by their horns (often spiraled or ringed) and by their choice of open, flat country for grazing. But there are exceptions, and some animals that are called antelope (such as the American pronghorn antelope) may be something else. Perhaps the one thing that all these animals have in common is that they are edible, and some are highly prized as both table fare and for sport hunting.

Some of the spiral-horned antelope of Africa, such as the eland, are quite close to cattle and grow to a large size. The so-called grazing antelope—24 species—live mostly in Africa, with a small distribution in Arabia. These include the reedbuck, waterbuck, hartebeests, wildebeests, impalas, and others. One of the most beautiful of the grazing antelope, the Arabian oryx (which at one time roamed the flatlands in Jordan, Syria, and Iraq) has been hunted for meat and trophy almost to extinction.

The gazelle and the dwarf antelope—another 30 species in the Antilopinae subfamily—can be found in Africa, across the Middle East, and into China. These include the springbuck, the blackbuck of India, and the graceful gerenuk. The duikers (subfamily Cephalophinae) are also small- to medium-sized antelope of the southern half of Africa.

The goat antelope (subfamily Caprinae) include 26 species, such as the delicious musk ox of the far north, the chiru, and the chamois, as well as the Barbary sheep, the wild goat (*Capra aegagrus*, now domesticated worldwide), the mountain goat, and the American bighorn sheep. Some of these are highly regarded as hunting trophies as well as for meat.

All of the different antelope can be eaten and most have, at one time or another, probably been hunted for food and sport. Primitive peoples hunting with bow and arrow—and even early rifle-bearing hunters on camel or horse—probably did not make a dent in the antelope populations; but modern firearms and high-powered, scope-mounted rifles have changed that, especially on the plains, where hunters are aided by jeeps and similar vehicles.

At one time, antelope were hunted extensively on the plains and steppes with the aid of trained cheetahs. The Arabs, Abyssinians, and Mongols hunted in this manner and Marco Polo reported the following: "Kubla Khan owns a large number of leopards trained to hunt, and lynxes which are taught to catch animals. He also has lions, far bigger than the lions of Babylon, with beautiful striped pelts of orange, black, and white. They are used to attack wild boar, wild oxen, bears, wild asses, deer, and roebuck." No doubt some of these animals were antelope, and it is doubtful that Marco Polo and Kubla Khan had a working knowledge of exact zoological classification.

(See also GAZELLE; GOATS; SHEEP.)

ants These insects live everywhere, from the tropics to the arctic regions. In general, they do better in the warmer areas. They all live in colonies and abide by strict divisions of labor, some being soldiers, and so on. There are more than 5,000 species, and one of the more interesting is the genus *Atta*, called the parasol ant or the leaf-cutting ant. These ants cut out pieces of leaves and hold them over their heads while they take them along the path to the entrance of their nest. Hence, the name parasol. But they don't eat the bits of leaves. They chew them into a mulch and deposit it on the floor of their underground chambers. Fungus grows on the mulch, and the ants eat the fungus.

Parasol ants are eaten by most of the Indians of the Amazon Basin and by some in Central America. Winged parasol ants are trapped just as they emerge from the nest at the beginning of the swarm. Usually, only the winged ants are eaten, as the others have sharp spines on their necks and heads. Some Indians eat the winged ants raw, others cooked. Still others pull off the heads, legs, and wings, then toast the bodies, which have been compared to crispy bacon.

The several species of honey ants (genus *Myrmecocystus*), which live in dry areas, collect the honey that is exuded by galls on oaks and other trees. The honey is stored in nests under the ground, but the ants don't build combs like bees. Instead, there is a special kind of ant called the replete, which stores the honey in its abdomen and, when full, hangs upside down from the ceiling of the compartment. Of course, the replete's abdomen swells up to grotesque proportions and the ant is helpless.

Honey ants can be found in the arid regions of the American West, where some of the local Indians dug into their nests and ate the repletes for taste and food value. Insect authority Ross E. Hutchins said that he tasted this ant honey near Pike's Peak and found it to be a little sour and not as palatable as honey from bees. But at least some of the Indians of Mexico and the southwestern United States were said to prefer the ant honey. The honey has also been widely eaten by the natives in Australia, who reportedly pop the replete's abdomens like grapes.

Other ants have been used in various kinds of medicines and cures. Ants with large biting jaws have even been used for stitching wounds in humans. To accomplish this, the ants were first placed in position to bite the skin. After the bite, in which the jaws apparently locked shut, the ant's body was pinched off, leaving the jaws intact. The process was repeated until the open wound was closed. This strange practice has been traced back to Hindu writings of 1000 B.C. It has also been used around the Mediterranean, and in South America.

(See also INSECTS.)

armadillos These heavily armored creatures are commonly used for food in South and Central America, and are not infrequently eaten in Mexico and the extreme southern parts of the United States. In Texas, they are jokingly called "possum on the half shell." The texture and flavor of the meat is usually considered to be very good. Although Charles Darwin said that it tasted like duck, it is more often compared to opossum, pork, and guinea pig.

Twenty species of armadillos now live in the Americas, mostly in the tropical zones. The common long-nosed or nine-banded armadillo (*Dasypus novemcinctus*) headed north some years ago, crossing the Rio Grande in the 1880s. This creature (like its cousins) doesn't float readily and must inflate its stomach and intestines with air before it can swim. Since it takes several minutes to pump itself up sufficiently, it will often walk across the bottom of small streams and ponds instead of swimming. In any case, the nine-banded armadillo has now reached Arkansas to the north and Argentina to the south. It feeds on insects, eggs, fallen fruits, fungi, and most anything it can find to eat on the ground or immediately under the surface. It has poor eyesight and finds its food mostly by a keen sense of smell. Armadillos tend to grunt and squeal a lot while feeding.

Some of the armadillos are too small to provide much meat, but the giant armadillo (*Priodontes maximus*) grows up to 5 feet long from snout to tail and can weigh as much as 130 pounds. In ancient times, an even larger armadillo lived in South America. It had a shell up to 10 feet long, which was used as roofs or tombs by the early Indians.

armored catfish (family *Loricariidae*) With some 400 species in several genera growing in Central and South America, these are different from the world's other 1,500 species of catfish in that they don't have skin or scales. Bony plates cover the entire body. The flesh of the armored catfish is quite good, but it is hard to get at. The bony plating is difficult to remove by skinning or scaling. Filleting is also difficult. Some Indians of the Amazon Basin merely toss the fish, whole, into the fire. After a while, they rake the fish out of the coals and crack open the shell to get at the meat.

The armored catfish can be caught by ordinary hook and line, but some of the species grow up to 100 pounds and are difficult to land. Even larger catfish grow in the Amazon—up to 500 pounds.

(See also CATFISH; FISH.)

aruaná (*Osteoglossum bicirrhosum*) These bony-mounted fish live in the Amazon watershed and prefer sluggish, weedy streams and shallow lakes. They grow up to 6 or 7 pounds. Because of their habit of swimming in schools near the surface of the water, they are relatively easy to catch with bow and arrow or spear. This makes aruaná easy prey for the natives, who use them for food. Not highly regarded as table fare, they are usually eaten in salted form. Aruaná eagerly strike artificial lures, especially if they are cast immediately in front of a school. They are, however, relatively difficult to hook because of their bony mouths. Once hooked, they are good fighters and make frequent jumps.

(See also FISH.)

Atlantic cutlassfish *(Trichiurus lepturus)* Despite its silvery skin, the cutlassfish is a thin, ugly creature with a wavy dorsal fin that stretches from head to tail. It resembles an eel with the oversized head of a many-toothed pike. The Atlantic cutlassfish is found in the Atlantic, the Western Pacific, and Indian oceans, as well as in the Gulf of Mexico. Its cousin, the Pacific cutlassfish, can be found from California to Peru. Both species prefer to live in bays and in the shallow areas of the ocean.

The cutlassfish is often caught on bait as well as on artificial lures. It grows to about 38 inches in length and attains a weight of about 2 pounds. Possibly because of its appearance, the cutlassfish is not widely accepted as a food fish in America, but it is eagerly eaten in Japan and a few other countries.

(See also FISH.)

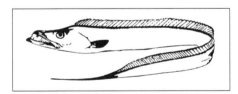

Atlantic cutlassfish

auks (family *Alcidae*) The 23 species of auks are seagoing birds that nest on barren shores and rocky cliffs around the fringes of the Arctic Ocean. They resemble the penguins of the Antarctic and, like penguins, are excellent swimmers that feed entirely in the sea from fish and aquatic matter. All of the extant auks have wings, and some loners even winter as far south as the Mediterranean Sea and Florida. In summer, however, they gather at their circumpolar rookeries in very large numbers.

The birds and their eggs are used for food by the peoples native to the coastal areas of the far north. The Eskimo of Greenland are especially fond of a relatively small auk called the dovekie, *Plautus alle.* (In New England, the dovekie is also known as the pineknot.) These birds nest in the cliffs, and are caught by nets as they fly by the Eskimo hunters. Although the dovekie is primarily a bird of the sea and the seashore, some of the inland Eskimo traveled a hundred miles or more for the summer harvest. Only a few years ago, the dovekie and other auks were caught by the thousands not only for immediate consumption but also for preservation. In order to store food for the coming winter, the Eskimo packed the birds, feathers and all, into seal skins. The skins were, of course, lined with blubber, which acted as a preservative. Birds and other meat preserved in blubber (called *giviak*) were highly prized as food.

The Eskimo also eat the large eggs from various auks and, at one time, these were also preserved for winter use. The yolks were packed, sausagelike, into long casings made from seal intestines. These were then hung in the shade to dry slowly.

Now extinct, the great auk (*Pinguinus impennis*) once populated the shores of the North Atlantic, from Newfoundland to Scandinavia. It was a flightless bird and was easily captured by early seamen. Since it made good food for consumption at sea, the birds were herded alive across a gangplank and onto ships. Reportedly, the last few of these strange birds were caught for food on a small island near Iceland in June 1844. The great auk was mentioned by early explorers, and modern archaeologists have found its remains in the kitchen middens of various prehistoric peoples of the far north. The bones of great auks have also been found in shell mounds of Denmark, dating back to the early part of the Neolithic Age. The birds were as big as modern-day geese. They had small wings, which were apparently used for swimming, not flying.

(See also BIRDS.)

B

babirusa (*Babyrousa babyrussa*) This wild pig of the Malay Archipelago is not merely edible; it is highly prized as food in the islands, where it grows up to about 200 pounds. The babirusa has no hair, and its rough gray hide, together with its squat shape, makes it resemble a rhinoceros. It has unusual tusks that do not form a part of the mouth. Instead, the tusks come out of the top of the upper jaw, jutting upward and back, almost touching the animal's forehead. These tusks have no apparent function. The babirusa feeds mostly at night on fruit and grass. It also roots for food in the soft dirt. Easily tamed, the animal is believed to be vulnerable in the wild, but it is not yet on the endangered species list.
(See also PIGS.)

badgers Nine species of badgers live in North America, Europe, Asia, and Africa. They eat small hole-dwelling animals such as chipmunks, which they catch by digging, but they also feed on fruit, roots, and worms. All of the badgers are edible, and have indeed been eaten by peasants during hard times in England and elsewhere. Only the tailless Malayan stink badger (*Mydaus javanensis*) of Southeast Asia (called *teledu*) is actively pursued by the natives for food. Several other species are hunted or trapped for their hide, and they are no doubt eaten on occasion. In the past, the badger's stiff hairs have been widely used for making paint and shaving brushes.

barnacles These crustaceans (related to crabs, shrimps, and lobsters) attach themselves to ships, buoys, cables, and other structures, thereby becoming a nuisance to man. Indeed, they can increase the fuel costs of ships by as much as 40%. Some species, especially the goose barnacles (*Lepas* species), are eaten in Chile, France, and elsewhere. They have the taste of crab or lobster, and are usually steamed or sautéed in butter.

barracudas The great barracuda, *Sphyraena barracuda*, reaches a length of about 6 feet and a weight of 100 pounds. It has been known to attack man, and is especially feared in the West Indies. Often simply called cuda, the barracuda can be found in the warmer parts of the Atlantic, from Florida to Brazil. Usually, the flesh of the barracuda is edible and is even highly praised in some quarters; but it can also be highly toxic to man. In fact, the great barracuda is first on the list of fish that carry ciguatera (a type of toxic poison). The others are all predatory fish of tropical waters. Some people believe that the larger barracuda is more likely to have ciguatera than the smaller ones, but this may or may not be accurate. To be safe, it is best to refrain from eating the flesh of the great barracuda.

The guaguanche (*Sphyraena guachancho*) is also a barracuda that inhabits the Atlantic, and goes a little farther north. It is a smaller fish and is not known to carry ciguatera. In fact, the guaguanche is one of the favorite food fishes in the West Indies. Because this species is easily confused with the great barracuda and lives in the same area, novice anglers should be certain of their catch before eating it.

The Pacific or California barracuda (*Sphyraena argentea*), on the other hand, is not a tropical fish and is safe to eat. Found from Alaska to Baja California, it is an important commercial fish, and is also popular with anglers. The flesh is very good, and its roe are highly prized. These fish are much smaller than the great barracuda.

The northern sennet (*Sphyraena borealis*) is a smaller cousin of the barracuda. It can be found as far north as New England and south into the tropics, where it mixes with the southern sennet, which lives as far south as Brazil. The sennets are very good eating and are not known to carry ciguatera. They are sold commercially in the tropical regions. They are relatively small fish, up to 15 inches long, and are not much esteemed as game fish, although they are caught by anglers on light tackle.

barramundi (*Lates calcarifer*) This large game fish can be found from northern Australia, around the coast of India, and on to the Persian Gulf. The barramunda is a catadromous fish, growing to adult size in streams and then moving downstream to spawn in brackish water mud flats. The fish grow quite large; indeed, catches of individual barramunda weighing over 500 pounds have been reported. The barramunda tends to hide in heavy cover and ambush its prey, much like a largemouth bass. This habit makes it an excellent game fish, and it will take all manner of artificial lures, including a rubber barra frog. The fish is highly rated as table fare, and it fetches high prices at the market.

(See also FISH.)

bass The most popular sport fish in the United States, the largemouth bass (*Micropterus salmoides*), is one of six species of black bass. It really belongs to the sunfish, not the bass, family. Such bass also have local names. In Louisiana, for example, the largemouth is often called green trout. In rural Alabama, the name trout is often used. However, in recent years, the popularity of bass clubs and tournaments—some of which have purses of over $200,000—has helped establish the term "bass" almost everywhere. These tournaments and clubs grew from the development of the bass boat, rigged with foot-controlled electric fishing motors and sonar devices, and from the building of numerous large impoundments on American rivers. Although the largemouth is very tasty, it is strictly a sport fish these days and is not sold commercially. Many anglers practice the catch-and-release policy.

The black bass also include the smallmouth bass (bronzeback), the Florida largemouth (which grows bigger than the regular largemouth), the spotted bass, the redeye bass, the Guadaloupe bass, and the Suwannee bass, which lives in the Suwannee River system of Florida and has a blue belly. All of these fish make very good eating and are also excellent sport fish. The black bass—and especially the largemouth—are very aggressive and will try to eat anything of reasonable size that moves through the water. Movement is the key. The black bass feed mostly by sound, whereas trout feed by sight and catfish feed by smell. Black bass can be caught on flies of suitable size and on a variety of artificial lures.

Unfortunately, many other fresh- and saltwater fish are sometimes called bass. The small white bass, which thrive in great numbers in large impoundments, and the large striped bass (rockfish or striper) come readily to mind. Both of these are popular sport fish, and are highly regarded as food. There is also a hybrid bass—a cross between the white bass and the striped bass—which is called sunshine bass in Florida.

The real bass is the sea bass, *Morone labrax*, a popular European game fish, which should not be confused with the black sea bass, a bottom feeder that is popular along the Atlantic coast of the United States. (According to Waverley Root, the black sea bass is served at Chinese banquets with its head pointed toward the guest of honor.) Excellent eating, the sea bass is also called gray bass and salmon bass. The giant sea bass, or jewfish, also makes good eating and grows up to 700 pounds.

Many other fish are also called bass, and listing them would take up far too much space. For example, the tripletail is sometimes called buoy bass. Also edible are the rock bass, the yellow bass, and the sand bass. The now-famous blackened redfish, a spicy Cajun way of cooking fish fillets, has become something of a fad in the 1980s and '90s. This is actually made from the red drum, or channel bass.

(See also FISH.)

bats Except for the polar regions, high mountaintops, and a few remote islands, bats of one species or another live all around the world in very large numbers. In spite of population declines in recent years, an estimated one fourth of all mammals on earth are bats. Most of these flying mammals go unseen by man simply because they are creatures of the night and roost in caves or otherwise stay hidden during the day. There are some 951 species of bats. These range in size

from the nearly extinct 0.052-ounce Kitti's hog-nosed bat to the Samoan flying fox, which has a wingspan of 6½ feet.

Some species of bats have learned to live in barns, house lofts, and other structures built by man. Most bats, however, inhabit caves and natural hiding places, and have therefore suffered greatly from the sprawl of civilization and the clearing of forests. On the whole, bats are declining in numbers year by year, although millions of some species may still be seen in one colony. In European countries and in some other places, bats are protected by law.

The good deeds performed by Batman of comic strip fame notwithstanding, the name of bats has suffered in the Western world from association, in art and symbolism, with vampires. In medieval Europe, the devil himself was sometimes depicted with bat wings. In the East, however, especially in China, the bat has long been a symbol of happiness and longevity; it was also regarded as such in ancient Persia.

Like most other mammals, bats are eaten in various parts of the world. The large flying foxes, which weigh up to 3 pounds, are highly esteemed as food in Africa and Australia. Bats in general are also popular in parts of China and Asia. Although the bat has been hunted extensively in some areas, natives armed with a spear or bow and arrow can take only a few, although large numbers may cluster in one spot. Modern shotguns, however, can kill or wound 20 or 30 with one shot. In Guam, the flying fox *Pteropus tokudae* has become extinct in recent years, partly because of meat hunting. The meat of other species of flying fox is sometimes brought to Guam from Yap and other islands. Some bats that live by day in caves with a narrow opening can be captured in very large numbers by meat hunters armed with fishing nets. Interestingly, a large fruit bat in the Philippines is hunted not only for its meat, but also for its fur. (All bats, being mammals, are covered with fine hair instead of feathers.)

In the past, bats have been important sources of fertilizer for people in many parts of the world. Most of the guano comes from cave bats and is, of course, nothing more than the droppings that have accumulated on the cave bottom. This layer of guano can be quite deep, and some caves have been mined for hundreds of years. In Colorado, more than 2,000 tons of guano have been removed from a single cave. In this region at one time, as many as 50 million bats belonged to a single colony. The Carlsbad caves of New Mexico have been mined for guano, which is said to have been used during the Civil War as a source of nitrates for use in making gunpowder.

bears The meat of most bears is very good or very bad, depending partly on how the animal is handled. Most bears are large animals and therefore contain lots of body heat. They also have thick fur, which keeps in the body heat. Consequently, quick and proper field dressing is required if the meat is to be prime. While all bear species are edible, there is a big difference among individuals. Speaking of the American black bear (*Ursus americanus*), Marjorie Kinnan Rawlings said, "A male bear in the mating season, like a boar hog, is not fit to eat. A female nursing bear not only has tough and stringy meat, but for humanitarian reasons should never be destroyed. A young male

bear in the off-season provides meat better than the best beef."

In the Arctic, the huge polar bears (*Ursus maritimus*), which can weigh up to 1,500 pounds, have been important to some Eskimo and other peoples of the arctic regions not only for food but also for skins. Although the meat is good, some authorities caution that sometimes the liver of a polar bear can contain such a high concentration of vitamin A that it is toxic to humans. Hunted extensively for sport, food, and hide, the polar bear has been greatly reduced in numbers in recent years and is now considered to be a vulnerable species. It is still hunted on a highly regulated and controlled basis by some Eskimo and other natives for subsistence only. The conservation and management efforts are international, involving the United States, Canada, Greenland, Denmark, and Russia.

The grizzly bear (*Ursus arctos*) is widely distributed in the northwest part of North America and from Europe eastward to the Pacific coast of Siberia. It is also native to the Pyrenees, the Alps, and the Caucasus mountains. The grizzly bear was once an important source of meat in northern Europe. In Russia, bear hams are cured just like pork hams and are delicious. In Azerbaijan, a kebab called *shashlik* was at one time made with wild bear and other game. In Poland, bear meat was used in early recipes for *bigos*, a dish that was traditionally served following a hunt.

The grizzly has been a very important part of the diet of peoples who live in some areas where salmon, a favorite food of the bears, run upstream to spawn, as in Alaska, parts of Siberia, and some of the islands of northern Japan, where the Ainu have cultural ceremonies in which they drink bear blood. The prehistoric cave people are also believed to have had cultural and ritualistic ties to the bear, which, in Europe, was probably the grizzly or its forefathers.

The grizzly is also called the brown bear, and there are several subspecies, such as the Kodiak bear (*Ursus arctos middendorffi*). The American black bear, mentioned above, lives in much of North America. It is eaten wherever hunting is legal, and it is often caught by poachers who sell the meat through the black market. The Asian black bear, sometimes called the Himalayan black bear, inhabits the forests and brush country from Iran to Japan. The other species (the sloth bear, the spectacled bear, and the sun bear) are not widely distributed.

Bear meat, like pork, can carry trichinosis and should be cooked well done. All parts of the bear are edible (except possibly for polar bear liver), and the paws have long been considered a culinary delight. In China, it has been said, the hooves of water buffalo have been cooked and served as bear paws, which, of course, are usually in short supply for modern man. Bear fat was highly prized as lard in early American cooking, and it makes excellent cracklings.

beavers The North American beaver (*Castor canadensis*) lives along and in streams and lakes from Alaska to the Gulf of Mexico. The felt hat, made with beaver fur, became something of a fad during the 18th and early 19th centuries, and the great demand for the pelts accelerated the exploration of North America. The beaver, which weighs up to 66 pounds, is not widely eaten in the United States, but it is much more appreciated in Canada, where the meat is a source of income for some Indian tribes. In its prime, beaver is a mild meat and has been compared to veal. The American Indian and the mountain men ate beaver, and are said to have considered the large paddlelike tail to be the best part, especially when roasted in the embers of a fire.

At one time, the beaver was almost wiped out from much of its original range. In recent years, however, it has made a dramatic comeback in many areas and is considered to be a pest in some timber producing regions. Its habit of damming up streams, resulting in impoundments that kill standing trees, together with its habit of gnawing the bark off trees on dry land, puts the animal at odds with timber growers. In order to control the rodents, many states have very liberal hunting and trapping laws, but this doesn't help in the warmer parts of the United States, where the pelts aren't worth much and where the meat isn't well received. The U.S. Department of Agriculture has hired animal control experts to keep the populations down, and large timber companies have either offered bounties or hired trappers. One company even developed a beaver birth control chemical, which worked but was never manufactured because wild beavers refused to eat the bait. In the northern parts of the United States and Canada, the beaver pelt is more valuable and control is not as difficult.

The European beaver (*Castor fiber*) was once plentiful throughout Eurasia, but it has become rare or even extinct in much of its former range. The North American beaver has been stocked in Finland, Poland, and parts of Russia.

Both the North American and the European beaver belong to the same family. Another rodent, the mountain beaver (*Aplodontia rufa*) is the only surviving member of a separate family. Its range is limited to a narrow band along the Pacific Coast of the United States and Canada. The mountain beaver grows to only 3 pounds. The other beavers, by comparison, reach 60 pounds or better, and are the world's second largest rodent. Because the animals (like fish) are seldom seen out of water, they were at one time considered to be Lenten fare.

bees and honey Some 20,000 species of bees and beelike insects live in various parts of the world, and they are usually plentiful wherever flowers grow. Some of the bees, entirely dependent on pollen and nectar of flowers, gather and store honey. These bees—known as honeybees—are of great importance to man. There are four species of honeybees. The most common, which has been widely domesticated, is *Apis mellifera*. In the wild state, its original range was Europe, Asia Minor, and Africa. The other species of honeybees are native to India and Southeast Asia, where they live in the jungles as well as in agricultural areas. In some places, bees tend to make their hives in the rocks of cliffs instead of in trees.

Of course, getting honey from wild bees has always been practiced by man as well as by bears. Wild honey was no doubt the first sweetener that was enjoyed by man, and even today many people take advantage of wild honey whenever they find it. (Some knowledgeable people even hunt wild honey by following the direction in which the local bees are flying.)

Whether or not the bee can be called domesticated, it has clearly been controlled or manipulated by man, and beekeeping is the oldest form of animal husbandry. At first, bees were kept in hives made of hollow logs or cylindrical hives made of woven straw and other materials. These days, highly efficient multiple-story bee hives have been developed, and they are moved from one location to another on large trucks (or barges), often for long distances, to follow the bloom of clover, orange blossom, or other plants. In addition to producing large quantities of honey, modern beekeeping is important to pollinating some crops; in many cases beekeepers are even paid by farmers.

Honey has been used since recorded history as a sweetener, and from it the first fermented drinks were made. Mead, for example, is a very old and almost universal drink that was sacred to the ancient Druids. The word "mead" comes from the Sanskrit *mahdu*, indicating an ancient origin. At one time, Great Britain was called the Honey Isle of Beli, because the peoples drank lots of mead. In France, the custom developed in which newlyweds partook of a honey drink for a certain number of days—hence the term honeymoon or honeymonth. In East Africa, honey is mixed with water and allowed to ferment, making a wine or beer called *tetsch*. Vinegar can also be made from honey.

Honey was often mixed with other foodstuffs such as milk, cheese, bread, cereals, and meat. The ancient Assyrians, for example, ground and mixed protein-rich locusts with honey and dates, which was used as a spread on breads. Modern man doesn't realize that honey (as well as beeswax) is a natural preserver of all organic matter. In fact Alexander the Great was embalmed with a mixture of honey and beeswax. The Egyptians also used beeswax for embalming.

Of course, beeswax has been used in very large quantities for candles and other products, including many art objects made with the aid of wax. A sticky substance known as propolis is used by bees to seal the inside of their hives. Man uses it in varnish and other products.

In addition to honey, wax, and propolis, the honeybee itself can provide food for man. In China, honeybees are fried. In Burma, the grubs and eggs are used to make soup. Some primitive peoples ate the larvae of bees to relieve constipation. Some modern peoples are said to eat chocolate-covered baby bees purchased in gourmet food shops.

All manner of magical and curative powers have been attributed to honey, such as the long life of the peoples of the Caucasus. Some people, in the belief that it promotes long and healthy life, collect and eat the pollen that is brought back to the hive on the bees' legs. Greek athletes ate honey before the Olympic games, and Roman soldiers ate it to promote long life.

On the Jewish holiday Rosh Hashanah, honey is eaten with an apple to symbolize a sweet new year. Although most of the honey used by mankind has been gathered with the aid of bees, ants and other insects also gather honey that is occasionally used by man.

(See also ANTS; INSECTS.)

beetles There are some 250,000 species of beetles, which usually have four wings, a hard shell and a formidable appearance. They range in size from 1/100 of an inch long to 7 inches long (rhinoceros beetle, *Dynastes hercules*.) All of them are edible, and a few species have been eaten in large numbers. Usually, the larvae of beetles are more popular as food and are consumed—raw, fried, or roasted—by Africans, Austra-

lians, Frenchmen, and Americans. The beetles themselves are also eaten here and there. In China, the diving beetle (family Dystiscidae) is raised in freshwater ponds for food, and is considered to be a great delicacy. The desert Bedouins of Egypt eat dung beetles as a sort of coming-of-age rite for boys.

The larvae of beetles as well as the grubs of butterflies or moths were important as a source of protein for the aborigines of Australia. Some of the beetle larvae feed on fallen logs; the aborigines developed a keen sense of hearing and a method of tapping on logs to locate the larvae. The idea, of course, was to avoid wasting time by digging into logs in which there were no larvae.

(See also GRUBS AND WORMS; INSECTS.)

birds Most of the 10,000 to 12,000 species of birds are edible. Fresh bird eggs are also edible, in any stage of embryonic development, and make an excellent survival food. Apart from the domesticated species, the seabirds that nest by the millions in cliffs of islands or other remote land masses have been of more importance than land birds to some peoples, such as the Eskimo, who catch them for immediate consumption and for year-round use either canned, frozen, salted, or otherwise preserved. The large birds have also been important simply because they contain lots of meat. These include the wild turkey of the Americas, the ostrich of Africa, and the emu of Australia. Sometimes birds that flock, such as rice birds, have been valued locally for food, and have been netted as well as taken with shotguns and dynamite.

The upland game birds, such as quail and partridge, are important to hunters but are not really too significant as food, however tasty they might be. At one time, hunting such birds with the aid of falcons or other birds of prey was popular in Asia and in Arab lands; it is still practiced today in these places, as well as in Europe and the United States. The advent of the shotgun also helped make wing shooting a popular sport, and most wing shooters also want to eat whatever they bag.

Wild ducks and geese are hunted for sport in various parts of the world and are highly prized by some people as food. Many primitive peoples have hunted ducks, geese, and other birds with the aid of whirling sticks and bolas. The wild turkey is hunted in the United States, and it is very plentiful in some states, thanks to modern game conservation practices. Doves and pigeons are also hunted in the Americas, New Guinea, Egypt, and other places and are also very popular as table fare. Unfortunately, the passenger pigeon and some other birds were greatly reduced in number by market hunters and by loss of habitat. Most of the upland game birds feed on grain and grass seeds, but some such as woodcock and snipe also feed on worms and grubs. The wild turkey feeds on grain and acorns as well as on chufas, grubs, snails, and whatever else it can find. Ducks, of course, feed on fish and aquatic matter as well as on grain. The diet of these birds very often determines their flavor.

Birds of prey are edible and are eaten by some people in many lands, but nowhere are they especially noted as tablefare. Hawks, owls, eagles, and even buzzards have been eaten, however. Songbirds are eaten in most parts of the world, although killing them may or may not be legal. Robins, for example, are killed and eaten by many rural people in the United States although they are protected by law. In Europe, songbirds have always been popular as food among the peasants and the epicures, and some of these were highly prized by the Roman epicures, who fattened them for the table. Water birds and wading birds have been hunted for food and sometimes for sport. Storks, cranes, and flamingos have all been eaten in the past, and some are still eaten today. In the United States, sandhill cranes are now hunted legally in some of the plains states.

Of the domesticated birds, the chicken is by far the most important on a worldwide basis. The hundreds of varieties that have been developed all came from the red jungle fowl (*Gallus gallus*) of southern Asia. They were first domesticated in India or Burma before 2500 B.C., and were established in Europe by 1500 B.C. The domestic ducks all probably descended from the wild mallard. They are now grown for meat, eggs, and feathers in most parts of the world; in China they are perhaps more important as meat than chicken. Geese were probably the first domesticated poultry, but these days they are not as important as duck or chicken. The domestic turkey, of course, came from the American wild turkey, which itself probably originated in Mexico. Other birds have also been domesticated and are used for meat and eggs, including the pigeon, peacock, guinea fowl, and pheasant. Ostriches are also raised on farms for meat, eggs, and feathers.

These days, quail and other game birds are raised in captivity and released in prechosen spots for the hunter's benefit. Often, bird dogs point the location of the birds before they are flushed and shot. Ducks are also released and permitted to fly by hunter's blinds. At one time, pigeons were released from boxes as targets for wing shooters. The clay pigeon has replaced the live bird on skeet ranges.

Baiting and fattening more or less wild birds for the hunter's table is an old tradition, which goes far back into our history. According to Marco Polo,

> There is a valley near this town where the Great Khan has huge numbers of great partridges, called *cators,* reared under the supervision of guards. In the summer he has millet, panic grass, and other grain for the birds sown on the slopes. No one may harvest these crops because they are for the birds to feed on at will. In winter millet is scattered on the ground for them. The birds are so used to being fed in this way that they flock to the keeper as soon as he whistles. The great Khan has plenty of birds here, but he does not stay on the plain during the winter when they are good and plump because of the cold. Wherever he goes, however, he has large quantities of these fat birds sent to him on the backs of camels.

(See also ALBATROSS; AUKS; BIRDS' NESTS; BITTERNS; BLACKBIRDS; BUSTARDS; CAPERCAILLIE; CASSOWARIES; COOTS; CORMORANTS; EIDERS; EMUS; FLAMINGOS; 2GUINEA FOWL; GULLS; IBIS; JABIRUS; LIMPKINS; LOONS; ORTOLANS; OSTRICHES; PEACOCKS; PETRELS; PIGEONS AND DOVES; PRAIRIE CHICKENS; RHEAS; SHORT-TAILED SHEARWATERS; SWANS; THRUSHES; TROPIC BIRDS; TRUMPETERS; WOODCOCKS.)

birds' nests Because most birds' nests are made of twigs, there may be a question as to whether this entry belongs in the plant section of this book. However, the birds' nests referred to here, used in soups in China, Vietnam, and other areas, are made by the swifts of genus *Collocalia.* The main building material in the nests is the birds' gelatinous saliva—animal matter. In the finished nests, this building material is shaped like strings or noodles.

From the culinary viewpoint, the best nests are made with red strings, but most of them are a sort of translucent white. After the nest is taken from the wild, it is soaked in hot water for several hours. The "noodles" will start to unravel, at which time a little vegetable oil is put into the water. The nests are stirred. After a while, more hot water is added, which brings all oil and impurities to the top. The process is repeated several times. The noodles are then boiled in chicken or beef stock, along with rice, vermicelli, and lotus seeds. In addition to its culinary properties, the soup is also believed to be an aphrodisiac. The distinguishing ingredient—birds' nest—can be purchased in Hong Kong, Singapore, and other oriental markets. However, these nests are quite expensive these days.

The swifts nest in caves, often on islands, and gathering the nests is an ongoing business in some places. The islands off southern Vietnam, near Nha Trang, are said to be productive, and there are other local sources that are popular. Swifts' nests can be found from Madagascar across Indonesia and Australia and on into French Polynesia. Some books list the swallow's nest instead of the swift's nest. The birds look alike, but their nests are very different. The problem is in confusing the name of the two birds. In North America, the common chimney swallow, a migratory bird, is really a swift (*Chaetura pelagica*). It also uses saliva in its nest.

(See also BIRDS.)

bison Often called buffalo, the American bison (*Bison bison*) was a very impor-

PEMMICAN

The word *pemmican*, came from the Cree Indians and means "journey meat." The idea behind pemmican, of course, is to pack a maximum amount of nourishment into a minimum bulk. To make pemmican, the Indians hung thin strips of venison or buffalo in the sun for a few days. After the meat dried, they pounded it into a pulp and mixed it with the fat from a bear or a goose and with pulverized dried fruit. (Modern practitioners might use hog lard or any fat that doesn't require refrigeration.)

    1 pound (454 grams) dried meat or jerky
    1/2 pound (227 grams) fat that doesn't require refrigeration
    1/2 pound (227 grams) dried apricots (or similar dried fruit)
    salt and pepper to taste
    brine soak (see below)
    melted paraffin and cheesecloth

Cut the meat into thin strips, as when making jerky. Soak the meat for an hour in a brine made with 1 gallon of water (3.785 liters) and 1 cup (227 grams) of salt. Hang the strips in the open until it dries, or dry it in the oven at 200°F (93°C) for about 8 hours. Leave the oven door slightly ajar to let the moisture out. After the meat has dried, pound it along with the fruit. (A grinder or a food processor can also be used.) Mix the pulverized fruit and meat into the lard, adding a little salt and pepper. Form the mixture, a thick paste, into small bars about an inch in diameter and 3 or 4 inches (7.6 or 10 centimeters) long. Wrap each bar in cheesecloth and dip it quickly into melted paraffin. It is, of course, always best to store pemmican in a cool, dry place, but refrigeration isn't required. Even today, pemmican is a highly nutritious mix for camp or trail.

tant source of food for the Indians—especially for the nomadic tribes of the plains. The reddish meat, tender and sweet, is prized even in modern times because it is low in fat (as compared to modern beef) and is still tender. The bison is currently being raised for meat in the United States, at least on a limited basis; how important the animal will become as a source of food remains to be seen. In the past, a number of attempts have been made to cross bison with domestic cattle, producing the cattalo. While initial cross-breeding has been accomplished, the male of the hybrid is infertile.

In any case, the importance of the bison to the plains Indians is difficult to overestimate. Every part of the animal was used for food, shelter, and clothing, as well as for other items of work and play. For war, shields were made from the bull's neck hide. Powder flasks were made from the horns, arrowheads from the bones, and hoes from the shoulder blade. Leather tanning agents came from the brains, fat, and liver. They made glue from the hooves and hide, and soap from the leftover fat. Knives were made from bone. Rafts and boats were constructed from the hide, which was also used to cover tepees. Blankets and robes were fashioned from the hide with hair. Moccasins were also made from the hide. The stomachs were used to make cooking utensils and water bags, and so on. Even the dried dung (chips) was used for fuel, and some tribes even made dice from buffalo bones. All of the meat was used either fresh or dried. Nothing was wasted. Even the bone marrow was eaten.

As a rule, the first parts to be eaten from freshly killed buffalo were the offal. The favorite parts, however, were the tongue and the fatty meat of the hump.

When the white man arrived in America, the bison numbered as many as 60 million, ranging from well into Canada down into Mexico. The settlers also made use of the buffalo's meat and hide, and they bought them from the Indians, who were glad to swap them for guns and other items of trade. In time, the white market hunters took over the business and killed off the buffalo in disgraceful numbers. William F. "Buffalo Bill" Cody, for example, is reported to have killed 4,280 animals within 17 months. Cody killed the animals as feed for railroad construction crews, but even more were killed for lesser purposes. The tongue of buffalo became something of a culinary fad in the eastern states and in such cities as Chicago. Many animals were killed merely for the tongues, with the hide and meat left to rot on the western plains.

The animals were killed off almost to extinction and, of course, fences and railroads tended to disrupt their way of life and limit their habitat. Fortunately, small wild herds were left in Yellowstone Park, and in a small part of Canada. At present, the U.S. government maintains herds of about 12,000 on a sort of controlled management basis. As stated earlier, the buffalo is also being raised commercially on a limited basis.

The European bison (*Bison bonasus*), or wisent, was eaten by ancient man and was recorded by cave paintings.

Bison

These animals, unlike the plains buffalo of the Americas, preferred to live in the forests, feeding on acorns, marsh plants, and so on. Instead of being killed within a few years, however, they were gradually reduced in numbers, both by hunters and by loss of habitat. They actually became extinct in the wild, but have been reestablished from zoo stock.

(See also BUFFALO; CATTLE.)

bitterns (*Botaurus*) All 12 species of these birds live in the reeds and cattails around lakes, along streams, and in marshes. They have stripes or other markings that blend in with the surroundings. Instead of flushing when a source of potential danger approaches, they tend to hold, although, like the bobwhite quail, they will usually flush at the last moment. This isn't to say that the bird ducks down and hides. On the contrary, it usually stretches its beak and neck straight up, thereby better blending in with the reeds. The bittern will also turn slowly as the source of danger passes, therefore making use of its best side (usually the breast) for the purposes of camouflage. In windy weather, it will even sway slightly in the direction the reeds are bending.

Bitterns are birds of good size, standing about 2½ feet. One species or another nests in temperate wetlands in North America, South America, Eurasia, South Africa, and Australia. The several species of smaller bitterns also nest around the world. All have similar habits and all are edible.

In the past, the bittern was hunted for sport and meat, and was especially popular with falconers. Reportedly, they were a favorite food of England's Henry VIII. (See also BIRDS.)

bivalved mollusks The mollusk comprises as many as 100,000 species that live in fresh water, salt water, and on land. Mollusks include the squid, the octopus, and the snail—all of which are edible—as well as many lesser known creatures. The bivalved mollusks include the oyster, the clam, the mussel, and the scallop, all of which have two hard shells that are opened and closed by an edible muscle. All of the bivalved mollusks are currently used for food in many lands, and they are among the oldest known food for mankind.

Evidence of the widespread use of mollusks by man is clear in the shell mounds that have been found around the world. Some of these mounds are so large that they were once believed to be raised beaches, made by some natural phenomenon. Digging, however, has led to the conclusion that they are manmade—a sort of garbage pile made of shells and bones, along with occasional artifacts, such as stone axes and lance points. Study of the shell mounds indicates that various kinds of oysters, clams, and mussles were eaten in very large numbers. Such a mound in Denmark dates back to the early Neolithic Age. Other mounds have been found in Scotland, Brazil, California, the Aleutian Islands, and elsewhere.

Clearly, primitive peoples made good use of oysters and clams.

The American Indians also ate large numbers of bivalved mollusks, both inland and on the seashores. At Shell Point on Harkers Island of North Carolina's Outer Banks, the Indians left hugh mounds of oyster and clam shells and, some people believe, tried to use them to build a bridge to connect Shell Point to Shackleford Banks. Later, the white settlers used the shells to build roads, uncovering many artifacts in the process. Oyster shells have also been used extensively to build roads in Florida.

In addition to food, the Indians and other early peoples used the shells of the mollusks for making pearly buttons and other items. The Alakaluf of Chile and the Andaman Islanders used the shells of bivalved mollusks as woodworking tools. The Andaman Islanders also fashioned an adze blade from a small wedge-shaped bivalve (pinna). The adze was used to hollow out canoes from tree trunks. In addition to their use as tools and utensils, the shells of bivalved mollusks have also been used for making jewelry and inlays. Of course, the pearl has always been highly prized.

(See also OYSTERS; CLAMS; MUSSELS; OCTOPUSES AND SQUID; SCALLOPS; SNAILS.)

blackbirds What is usually called the blackbird in Europe is a specific member of the thrush family, *Turdus merula*, and is highly regarded as table fare. The same bird is also quite plentiful in Asia. In America, several species are called blackbirds, such as the beautiful red-winged blackbird, *Agelaius phoeniceus*. There is also a tendency to call any bird of black color a "blackbird," provided that it is small enough to require 4 and 20 for the pie.

Larger black birds are called crows, rooks, ravens, and so on. In any case, the smaller black birds are sometimes eaten, especially in a blackbird pie. Basically, these pies are prepared rather like the other famous meat pies of the British Isles.

In America, large flocks of black birds of one sort or another sometimes do considerable damage to the yield of grain fields, and often cause problems when large numbers roost in town squares. All of these birds can provide good eating. The fish and game departments of some states have liberal laws concerning these birds and have encouraged hunters to go after them; but usually they meet with little success.

In Corsica, the blackbird feeds heavily on berries that grow on the lower slopes of the mountains. These plump blackbirds have a very good reputation as food, and they are the main ingredient in a Corsican speciality, *pâté de merles*. The dish is prepared with juniper berries and a local spirit flavored with myrtle.

(See also BIRDS.)

blood Mankind has always consumed blood, even when practicing cannibalism. Of course, any time fresh meat is eaten it will have some blood in it, so that, like it or not, everybody who eats meat will of necessity have to eat some blood. But some peoples attempt to get rid of the blood by bleeding the animal at the moment of slaughter, and by soaking the meat in cool water before cooking it. The Jews are forbidden by Mosaic law to eat meat that wasn't first bled and then salted; Arabs have similar restrictions.

Before the Islamic law, however, Arabs ate a patty made of camel's hair and blood, which was cooked over an open fire. In more recent years, several peoples of Africa roasted coagulated blood over hot coals. And even today the peoples of the British Isles and a good many other places eat dishes known as blood pudding, blood sausage, and even blood soup.

In some cases, drinking or eating blood was a practical matter—a way to draw nourishment from domesticated animals without killing them. This concept was applied especially by peoples whose culture depended heavily on the horse for milk, meat, transportation, and war. Both the ancient Sarmatians and the Mongols drew blood from their horses and drank it. According to Marco Polo, the Mongol warriors, at least on some missions, each carried with them a string of 18 horses. This permitted the horseman to draw blood from a different horse every day, thereby saving time, reducing the load, and precluding the need for cooking fires—the smoke of which could be spotted for long distances on the plains.

Blood has also been drunk in modern times, and not only by small and secretive cults. In Victorian England, blood was believed to ward off tuberculosis, and even the ladies of the day were known to make regular visits to the local slaughter house to drink glasses of freshly let blood.

Bombay ducks   (*Harpodon nehereus*) These creatures, popular food of India, aren't ducks or even birds. They are lizard fish that live in the warm waters along the coast of India and the mouths of streams, especially the estuary of the Ganges. They sport a big head, large teeth, and a cylindrical body. Predators, their maximum length is about 16 inches.

In India, the fish is filleted, salted, and sun-dried. Bombay duck is often found on menus in Indian restaurants, where, more often than not, it is eaten as an appetizer to the main meal. Under the name *bombil*, the fish is also an inexpensive source of protein for the Indian family, where it is used as an ingredient in other dishes.

(See also FISH.)

bonefish (*Albula vulpes*) Referred to as *banana*, *ikondo*, and *kpole* in Africa, *macabi* in the West Indies and Mexico, *raton* in South America, and *o'io* in Hawaii, the good reputation of the bonefish among saltwater anglers is beyond dispute. Its value as food, however, is a matter of opinion. It inhabits all the world's tropical seas and is fished extensively on the salt flats around the Florida keys, where it is seldom eaten by most people and is, these days, usually released as soon as it is caught. Some authors say that the flesh is not of good flavor. Others praise it highly, saying that the flesh is firm, white, and nutlike in flavor, and that it is good when prepared a number of ways, including baked and smoked. The roe is considered to be a delicacy.

The bonefish grows to 20 pounds or better, and its reputation as a game fish comes from its ability to make long, fast runs. It can, in fact, zip off from 50 to 100 yards of line, which puts a severe test on the angler's equipment. In any case, the fish is very valuable as sport fish in some areas, especially in south Florida and the Bahamas, where expansive salt flats are available, and the catch-and-release ethic is encouraged. In other parts of the world, the fish is sometimes caught by commercial fishermen and marketed.

(See also FISH.)

bone marrow See BONES AND BONE MARROW.

bones and bone marrow. Primitive peoples the world over, including early man, have enjoyed bone marrow. In Lapland, marrow of reindeer is highly regarded, and in the New World arctic regions it is sometimes called Eskimo butter. In Africa, the bone marrow of the long-legged giraffe is considered to be a great delicacy. Bone marrow is also used in some highly developed culinary traditions. In France, for example, the marrow of calf backbones is highly regarded in some circles; called *amourette*, it is used in recipes that call for brains, which it resembles in texture. Cannibals have also enjoyed bone marrow in various parts of the world.

Bones themselves are sometimes eaten along with meat, and are a good source of calcium. Canned sardines or salmon, for example, contain fish bones, which have been softened by the cooking or canning process. The Indians of California pounded deer vertebrae and the bones of salmon, then stored them for later consumption. The nomads in the Sahara grind up bones and mix them in with their food. Chefs all over the world know that bones boiled in water make for excellent soup stocks.

bots See MAGGOTS AND BOTS.

brown trout (*Salmo trutta*) This native of the Old World has interested anglers and gourmets since ancient times. It is entirely possible that the art of fly-fishing was first practiced in a river in northern Greece or southern Yugoslavia, where a certain spotted fish frequently rose to gulp down a certain floating insect. According to Aelian, a 3rd-century A.D. Roman, some clever angler devised an artificial fly of similar appearance by tying wool yarn around the shank of a hook and fixing on wings made from bird feathers.

In any case, the brown trout has always been highly regarded by anglers. Its original range was from the Mediterranean basin to the Baltic Sea, over to Norway and Siberia. It currently thrives in the cold water streams of the United States, South America, parts of Asia, New Zealand, and Africa.

There are a number of subspecies of brown trout, including the Kura strain in the Caspian Sea, which spawns only once during its lifetime, lays up to 30,000 eggs, and attains weight of up to 112 pounds. A strange relative in Ireland, called the gillaroo, feeds on snails and has red meat. Another relative in Turkey is called *Allah Balik*— God's fish. All of the wild brown trout make excellent eating. A more or less domesticated brown trout is raised for the market these days, and a "put and take" brown trout is raised in ponds and then released into streams for anglers to catch. There are, of course, several other species of trout that are caught by the angler for food and sport. (See also FISH.)

buffalo Several large oxlike, cud-chewing animals are called buffalo. One of the most important of these animals at the present time is the water buffalo, *Bubalus arnce (bubalis)*. These still survive in the wild, mostly in protected game reserves and in some very remote areas of India, Assam, Nepal, and other parts of Asia. They are dangerous animals and will charge hunters. The bulls are large, weighing 2,500 pounds or better, and have massive horns.

The water buffalo was domesticated in ancient times and is still widely used in Asia as a beast of burden. Domesticated water buffalo are also widely used in South America, Australia, North Africa, and Europe. In some areas with lots of wild lands, such as Australia, the domesticated animals have escaped and have formed feral populations. In its original range in Asia, the water buffalo has also escaped back to the wild, so that it is sometimes difficult to determine which is which. Whether domesticated, wild, or feral, the water buffalo, as its name implies, is fond of water and likes to wallow, especially during the heat of the day. They are sensitive to heat, and will often use a cake of mud on their hide to protect them from the sun's rays. Obviously, they are extremely valuable animals for farmers to own in areas where wetlands are cultivated, as in Bangladesh.

The water buffalo is, of course, edible, but the wild specimens are in short supply and difficult to bag. The domestic ones are usually more important as draft animals and as a source of very rich milk and butter. In fact, the buffalo contributes about half of the milk that is consumed by the people of India. The milk is also important in parts of Africa, China, and the Philippines.

Another species, the African buffalo (*Synceros caffer*), includes the Cape buffalo, forest buffalo, and other variations. Unlike the water buffalo of Asia, these animals have never been domesticated. With the bulls weighing in at about 1,500 pounds, they are somewhat smaller than the water buffalo; but they are just as mean and are especially dangerous when wounded. At one time, these buffalo lived from one end of Africa to the other, but hunting for food and sport—as well as destruction of habitat—has greatly reduced their range.

(See also BISON; CATTLE.)

bulldozers (*Scyllarides latus*) According to a 1976 issue of *Alabama Conservation*, the bulldozer is a seafood delicacy that isn't often found in the market. It's a deepsea decapod about which little is known, except that its tail is good to eat. Bulldozers are ugly creatures that live on the bottom of the sea between 16 and 40 fathoms.

Bulldozers are caught by shrimp trawls in the Gulf of Mexico, and they can be found up the Atlantic coast as far as North Carolina. Sometimes they are found in seafood markets, and are even served in restaurants that specialize in seafoods. But most of them are eaten by the fishermen who catch them, partly because they are so good and partly because they are not taken in significant numbers by the trawls. They grow up to about 14 inches long and weigh approximately 2 pounds.

burbot (*Lota lota*) Also called cusk, ling, lawyer, ling-cod, and, often, eelpout, the burbot is the only freshwater species of the cod family. A cold water creature, it looks like an eel with the head of a catfish. It lives in the deep holes of lakes and sluggish streams, and it usually feeds at night. Its range includes the northern part of the United States and Canada, as well as northern Europe and Asia. In the Great Lakes, its average weight is only a pound or so, but it grows much larger farther north. In Siberia and Alaska, a related species grows 5 feet long and weighs up to 60 pounds.

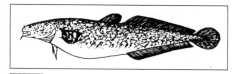

Burbot

Like the saltwater cod, the burbot has a large liver that is very high in vitamins A and D. Most Americans eat little or no fish liver of any sort (possibly because of the taste of cod liver oil), but the liver of burbot is considered to be a delicacy in Scandinavia. French connoisseurs poach it in white wine before making it into paté. In addition to the liver, the eggs or roe of burbot are edible and are prized in Finland and some other areas. In Finland, the roe is even more highly regarded than cavier from the sturgeon; the Finns use the burbot flesh to make a fish and potato soup called *madekeitto*.

Although the liver contains lots of oil, the flesh of the fish is mild and lean, similar to that of the cod. It should be more widely eaten in the United States and in Canada than it currently is, as it is plentiful in thousands of cold water lakes. Attempts have been made to make better use of the fish, but not with much success. An eelpout fishing tournament is held each year on Leech Lake in Minnesota, a state with lots of burbot habitat, as ice fishermen know. At one time, burbot was harvested in Minnesota for its liver oil, which was the major ingredient in a bottled dietary supplement.

(See also FISH.)

bustard This family of birds comprises 23 species. The best known is the giant bustard, *Otis tarda*. It has an 8-foot wingspan and weighs up to 32 pounds. The great bustard ranges from the grain fields of Europe to Manchuria. At one time, the great bustard lived on the steppes of

Russia and Siberia, but its numbers have been greatly reduced in recent years.

Other species range from Africa (16 species, including the kori, or giant bustard, which weighs up to 50 pounds and is the largest extant flying bird) to Korea and down to Australia. Often the Australian bustard is called a wild turkey.

All bustards are sharp-eyed birds of the plains. They are strong fliers, but usually they run from danger. In the past, they have been quite popular as a game bird, and were hunted extensively by falconers. The birds also make excellent table fare, having delicate and toothsome flesh; the little bustard (*Otis texrax*) is highly prized in Europe. Years ago, the bustard stood a fair chance against hunters on foot armed with bows and arrows, but today they are no match for hunters equipped with modern trucks and shotguns.

(See also BIRDS.)

C

cabezones (*Scorpaenichthys marmoratus*) A saltwater species, the cabezone weighs up to 25 pounds and inhabits the Pacific from British Columbia to Baja California. The cabezone is an ugly fish and changes colors—from green to cherry red—according to its environment. Often caught by anglers, the fish's flesh is very good according to some accounts, but is rated as only fair-to-good by others. A good deal depends on how the fish is cooked; frying it is not the best way to go. In any case, be warned that its green roe is poisonous to man.

(See also FISH.)

camels Many of the young American and British soldiers sent to Saudi Arabia in Operation Desert Storm were shocked to learn that bacon and pork chops are forbidden fare in that part of the world. Indeed, many of them had never fully realized that these popular supermarket meats, usually packaged in styrofoam and plastic wrap, came from a pig. The gastronomic shock went even further when they learned that some of the Saudis eat camel's meat.

Actually, the meat of camel has been highly regarded in some quarters. Aristotle praised it and Aristophanes said that it was food fit for royalty. Of course, the Roman gourmands also ate camel meat. The whole camel can be consumed, from the tongue to the tail. The hump of the camel is considered to be a prime cut; often it is cooked whole, like a roast. The stomach is also highly regarded. The camel's wide, soft foot, designed for walking in the sand, is considered to be one of the best parts of the animal for eating purposes. Camels' milk is also consumed by desert peoples.

The camel originated in North America some 40 million years ago. Before becoming extinct in its original range, it spread to South America and to Asia. Today, domesticated camels range from Morocco to Mongolia. The Arabian camel (*Camelus dromedarius*) has one hump, whereas the Bactrian camel (*Camelus bactrianus*) sports two. A few wild camels may be found here and there in remote regions of Australia and elsewhere, but these are really animals that have strayed from domesticity. The U.S. Army imported some camels in the late 1850s, when Jefferson Davis was secretary of war, for use in desert regions, especially for mail service. For one reason or another, the camels didn't work out and they were turned loose in the Southwest. The British have used camels, with some success, as desert cavalry.

In biblical times, camels were considered an important source of wealth and were kept in large herds. Moses forbade the Israelites to eat the meat, but the milk was commonly used for food. Clothing was sometimes made of camels' hair and was usually worn by poor or ascetic people. Such clothing was said to be durable and easy to keep clean; further, it did not hold perspiration or harbor vermin.

The Arabs have always eaten camels, especially on a survival basis. The Iranians, the Kurds, and other eastern peoples do not normally choose to eat the meat on a regular basis. Nomadic peoples of the desert lands, from North Africa to Mongolia, have always made very good use of the camel as a beast of burden as well as for meat. According to Marco Polo, camels were a staple in Madagascar during the Middle Ages: "An unbelievable number of camels are killed for their meat, which is eaten every day all the year around because it is thought to be the best and most wholesome meat in the world."

cane rats Two large species of cane rats live in Africa, south of the Sahara. The largest, *Thryonomys swinderianus*, can weigh as much as 20 pounds with a body length of 2 feet; its cousin, *Thryonomys gregorianus*, weighs up to 15 pounds and is almost as long. Both species live near the water, preferably in marshes or in thick reeds along lakes and rivers. They feed entirely on grass and other vegetation.

The cane rats often do considerable damage to crops and are generally considered to be pests. They are often killed for food as well as for pest control. (See also RODENTS.)

capercaillie (*Tetrao urogallus*) The bones of this large game bird—a grouse that may may weigh up to 12 pounds—have been found in kitchen middens of Denmark and other places. By modern standards, it is perhaps the best tasting of all the grouse, and perhaps of all the European game birds. Its delicate meat is whiter than pheasant. The bird can be found here and there, from Lapland to Turkistan, and south to the northern edges of Spain. The bird always lives in or near pine forests, which provide its food in winter.

In Scotland, the bird is hunted with the aid of beaters, who frighten the birds out of the forest and across the moors, where hunters await them. In other parts of Europe, the birds are sometimes taken by stalk hunting during the courtship season. Usually, the male capercallie has

a sharp eye and is difficult to stalk, but when courting it has a habit of sitting in the very top of a tree, stretching its neck up, and emitting a loud mating call. During this call, it closes its eyes. Thus, the hunter can advance by starts and stops until he gets within easy shooting range. (See also BIRDS.)

capybaras (*Hydrochaerus hydrochaeris*) These 140-pound tailless animals, which live east of the Andes in South America (from Panama to Argentina), are the largest of the extant rodents. Sometimes called water pigs or Orinoco hogs, capybaras always live near water. They dive and swim for pleasure, and their dens are likely to be under cover or under the banks along the water's edge. They can live in forests, though are more often found in wet grasslands, and prefer to feed on short grass.

Although they have been hunted commercially for food and leather, capybaras are still plentiful in some areas, such as Venezuela, where hunters are permitted to harvested them on a controlled basis. In other countries they are not plentiful and are protected by law. According to *The Encyclopedia of Mammals*, the meat of the capybara has been in demand in Venezuela ever since Roman Catholic missionary monks classified it as Lenten fare.

There is a move afoot in Venezuela to farm the animals. They can be herded by horsemen, and capybara farming may prove to be a profitable sideline for cattle ranches with suitable wet acreage. (See also RODENTS.)

carp (*Cyprinus carpio*) This large minnow, growing up to 80 pounds, is not commonly eaten in America, or at least not in the United States, where, ironically, it was stocked as a food fish in 1876 by the U.S. Fish Commission. Yet the carp is highly prized as food in Europe and Asia. Adding to the paradox, the carp is also a great game fish in Europe, considered to be exceptionally wary and hard to catch. In America, however, the carp is considered to be useless not only as food but also as sport. A few Americans hunt the carp for sport, armed with spears and bowfishing gear; this practice is encouraged by the various state fish and game departments as a means of holding down the populations.

As a rule, the fish is best when steamed or poached and is not suitable for frying. (*Larousse Gastronomique* allows deep frying very small carp.) Further, it is best when very fresh, and indeed it is often kept alive until needed. The carp should be skinned and the dark meat, which is tough and tastes bad, should be removed before the fish is cooked. If properly prepared, the carp is excellent when smoked. In addition, the fish is used in making gefilte fish, a traditional Jewish holiday dish. The roe is also edible and is sometimes made into caviar.

Although the carp is not valued in America, it is still the most important freshwater fish in the world and is raised for food in Europe and Asia. In some regions, carp are raised in connection with rice farming, and can indeed thrive in sluggish, muddy waters filled with vegetation. (In some areas, carp from muddy water are kept in clear water for several days before they are eaten.) In Israel, carp ponds are common in the kibbutzim. In short, the carp has been farmed for food since ancient times, and and some 1,500 varieties have been developed.

The crucian carp (*Carassius carassius*) is a different fish that is much smaller than the carp. It is of some value to anglers in Europe, and is also raised commercially. Although the flesh is good, such carp are not widely eaten except in Germany. Some of these fish are of ornamental value, and the goldfish (*Carassius auratus*) is a close cousin. (See also FISH.)

cassowaries These large flightless birds live in New Guinea, northern Australia, and on a few small islands in the region. Although they are smaller than the ostrich and the emu, they are nevertheless dangerous to man. Cassowaries are large birds, standing up to 5 feet tall, and they have a way of jumping up and stabbing with their very powerful legs. Their inner toe is long, sharp, and daggerlike. The cassowary is edible, and the birds are hunted for both meat and sport in New Guinea.

The birds have an impressive helmet, and the common cassowary (*Casuarius casuarius*) is especially striking, with two red wattles under a blue throat. The birds are fast runners, going up to 30 miles per hour, and are also good swimmers. Along the coast, they will even take to salt water. Bennett's cassowary (*Casuarius bennetti*) of New Zealand is raised for profit, partly for its valuable feathers, and also because the young birds are said to be rather like pets that run around loose like dogs. The older birds are confined to cages and fed garbage. The feathers are plucked for the market without killing the birds. (See also BIRDS.)

catfish Over 2,000 species of catfish live in the world's fresh waters, and a few more live in salt water. All of them are edible, and many species are eaten in various parts of the world. They are especially popular in the southern United States, while very large catfish are highly esteemed as table fare in parts of South America.

One of the best catfish for eating purposes is the channel catfish (*Ictalurus punctatus*), which lives wild in streams and is often caught in the swift waters below hydroelectric dams on rivers. They are marketed by a few commercial fishermen, who take them in baskets (wire or wood-slat traps) as well as on trotlines. Even more are marketed by catfish farms, which stock fry or fingerlings in ponds and feed them until they are ready for market. This has become an important industry in some states, such as Arkansas and Mississippi. Other species of catfish are also raised in ponds, such as the delicious white catfish (*Ictalurus catus*). At Auburn University, a catfish that grows much faster than normal has been developed by genetic engineering, but it has not yet been let loose from its carefully fenced ponds.

Many of the world's catfish are native to South America, and especially to the Amazon Basin. Here they are caught commercially by one means or another and are important to the economy of several villages. Some of these Amazon catfish, such as the pirarra and the pirarucú, grow up to 500 pounds. In certain rivers of Europe, very large catfish grow up to 600 pounds. One plant-eating catfish (*Pangasianodom*) grows to 6½ feet long in the Mekong River; in Thailand, similar fish have been raised in ponds and fed fruit and vegetable matter.

Only two kinds of catfish live in salt water. The gafftopsail (*Bagre marinus* and related species) grows to about 2 feet in length and is a good fish for the table, although it is not eaten in some areas. It often stays in brackish water. The sea catfish, or hardhead (*Arius felis* and related species), live in temperate and tropical coastal regions around the world, but they do not often stay in brackish water. They are edible but are not as highly esteemed as the gafftopsail. All of the seagoing catfish can inject a poison through their spine, and should be handled with caution. The same warning also applies to most of the other catfish. Even the tiny madtom can inflict a painful wound. One species of catfish in Africa can even produce 450 volts of electricity .

(See also ARMORED CATFISH; FISH; PIRARRA; PIRARUCÚ.)

cats The common house cat, *Felis catus*, was first domesticated from an African wildcat (*Felis libyca*) by the ancient Egyptians. During the 5th and 6th dynasties (2500 to 2300 B.C.), the domesticated cat was held to be sacred and was an important part of Egyptian life. Thousands of cat mummies—and even some mice mummies—have been found in ancient tombs. Slowly, the cat spread to other parts of the world. Instead of being sacred, however, the cat has more often been associated with sorcery and bad luck—especially black cats.

In any case, domestic cats are edible and are used as food during times of famine and siege. Reportedly, they have a flavor halfway between a rabbit and a hare. According to *Larousse Gastronomique*, rumor once had it that cats were used in some commercial kitchens for the preparation of rabbit fricassees.

In America, the meat of the bobcat was highly esteemed by the Indians and some of the pioneers. A few writers on the subject even proclaimed it to be the best of all wild game. Most modern hunters, however, hold rather strong opposing sentiments. A firsthand account was given by the late Euell Gibbons, who killed and ate a bobcat in the American Southwest. The "bob" part was quite tasty, he said, but the "cat" part was difficult to swallow. The lynx was also widely eaten in the northern parts of the New World, and might be a little easier for modern man to swallow. The lynx is still greatly valued as table fare in parts of Canada and Alaska.

Both lions and tigers have been eaten in various parts of the world, as have the cougar or mountain lion (also called painter, Florida panther, and so on.)

cattle All of today's cattle probably developed from a subspecies of the now extinct auroch or urus (*Bos primigenius*). These animals originated in Asia, then spread into Europe and North Africa. They were hunted for food by primitive man, and the Stone Age peoples painted the auroch on the walls at the caves at Lascaux, France. The last one died in 1627. These wild cattle were probably the first animals that were domesticated for agricultural purposes. Evidence of the domestication of oxen has been found in the Swiss lake dwellings of the Neolithic Age, and domesticated cattle also existed in Egypt and Babylonia, going back further than 3500 B.C. After domestication, cattle were used for food and milk and as beasts of burden.

Today, most cattle have been selectively bred for various parts of the world and for various uses. The Holstein and the Jersey, for example, have been bred for producing milk, while the black Angus was bred primarily for beef. Some breeds, such as the Red Poll, were bred to produce both milk and beef. The Texas longhorn came from a line of cattle that was brought to America by the Spanish, and it was suited for the sparse grazing on the semiarid lands of the American Southwest. The humped Brahman or zebu is from India, and has been used extensively in the southeastern United States to breed cattle that will do well in hot, humid conditions. In its native India, the zebu is used mostly for milk and as a beast of burden, but the flesh is not eaten. The Italian Chianina is the oldest, largest, and heaviest breed of cattle, dating back to the Tuscans, who bred it for sacrificial purposes.

Although a present move to produce lean, low-fat meat may become an important force in the cattle industry, the trend in the past has been toward producing a meat with lots of fat in the tissue. Most cattle in the United States is bred on the range or in pastures, and is then confined for fattening before it is sold. This practice is carried to the extreme in Japan, where the so-called Kobe cattle are fattened on beer and massaged in such a way that the meat is more marbled with fat.

The feral cattle that live scattered around the world include the herd of wild cattle in Chillingham Park in England. Cousins to the domestic cattle still grow wild in the world, although most of them are declining in number. These animals, all of which are edible, include: the Banteng (*Bos javanicus*), which lives in parts of Southeast Asia; the gaur (*Bos gaurus*), found in parts of India and elsewhere; and the endangered kouprey or

Cambodian forest ox (*Bos sauveli*), found only in a few areas of Indochina. True cattle also include the yak of Tibet. The latter animal has also been domesticated, and in fact has been crossed with regular cattle. Other wild animals that are closely related to cattle include the nilgai, American bison, the water buffalo, and the large spiral-horned antelopes, such as the eland.

(See also BISON; BUFFALO; YAKS.)

caviar See EGGS; FISH.

cavies (family Caviidae) These South American rodents, made of up 14 species, are all edible and some are considered to be delicacies by the local peoples. The cavies include the guinea pigs, which were being raised for food in Peru when the Spanish arrived in the 1530s. Most of the other species of cavies are very similar to guinea pigs in appearance. One of these, the rock cavy (*Kerodon rupestris*), is hunted extensively.

One species of the mara cavy called the Patagonian hare (*Dolichotis patagonum*) is much larger than the others, growing up to about 18 pounds. It looks like a large rabbit. Having long legs, it can run up to 28 miles an hour over long distances. A related species, the salt desert cavy (*Dolichotis salinicolum*), may no longer exist in the wild.

(See also GUINEA PIGS; RODENTS.)

chamois (*Rupicapra rupicapra*) Highly prized as venison, these sure-footed goat antelopes live in the mountains of Europe—Spain, Italy, Turkey, Austria, Switzerland, and the Balkans—and have been transplanted to New Zealand. Chamois are quite wary and nimble, even in rocky mountain sides. This ability to move surely and quickly among

the rocks helps them survive hunters as well as the wolves and lynx that share the same mountain ranges. Nevertheless, they are now rare in much of their original range.

In summer, the chamois feeds on mountain herbs and flowers. During this time it stays in loose flocks, except for the old males, which are rather solitary. In winter, the flocks split up as the food becomes more scarce, at which time the animals feed on lichens, patches of grass, and pine shoots.

Chamois hunting is considered to be quite a sport in some areas. It isn't a large trophy, as it stands only about 2½ feet tall at the shoulder. Nor is its rack impressive. But it is considered to be an exceptionally wary target, difficult to reach and difficult to stalk. In times past, the animal was the source of the popular chamois "cloth" used for buffing. Originally, the skin of the chamois was made soft and pliable by treating it with fish oil. Now, of course, chamois cloth is usually made from other skins (usually sheep or goats) or from specially treated cotton.

(See also ANTELOPE.)

Chamois

chimeras About 20 species of these sea fish live in cold, deep waters. Some species have been found as deep as 8,000 feet. They often hug the bottom, and are very poor swimmers, relying only on their fins instead of body movements. In fact, chimeras usually move by flapping only their two pectoral fins, which gives them an awkward up-and-down start-and-stop, or jerky movement through the water. (Some species also use their tail and dorsal fins.) The animals are somewhat bizzare in appearance, and derive their name from Greek mythology (i.e., a she-monster with a lion's head, goat's body, and serpent's tail). Most chimeras have a long hollow dorsal fin that is connected to a poison sac, and can be used accordingly. All of their teeth are fused together, designed to form a beak for cracking shells.

The flesh of chimeras is firm and good, but it is seldom used for food in most places. In China and New Zealand, however, the meat is consumed on a regular basis. In Scandinavia, the chimera is caught mostly for its liver oil, which is used in medicine and as a lubricant. A species called the ratfish (*Hydrolagus colliei*) living in the cold waters of the Pacific is said to produce an excellent gun oil.

chimpanzees (*Pan troglodytes* and *Pan paniscus*) These apes of central and West Africa feed largely on plants and fruits, eating up to 300 species at various seasons of the year. Although insects make up the bulk of their nonvegetable diet, chimpanzees will eat birds and mammals as well as eggs. If the conditions are right, they will even hunt and prey on pigs, antelope, monkeys, and even baboons. They have also been known to cannibalize young chimpanzees, usually from another tribe. Highly intelligent, chimpanzees use tools when gathering and eating food as, for example, when enlarging a hole with a stick in order to get to a nest of ants or termites.

The chimpanzee is considered to be excellent meat for humans, and is no doubt still eaten in some parts of their range, mostly Sierra Leone and eastern Zaire. In some other parts of their range, chimpanzees are more or less protected by local custom. The animals here may be quite tame and often visit man's fields and even the markets, no doubt in quest of easy food. In the wild, chimpanzees grow to about 150 pounds. The pygmy chimpanzee or bonobo, *Pan paniscus*, is a little smaller than the common chimpanzee, but there is also a good deal of difference between individuals of both species.
(See also MONKEYS.)

chinchillas and viscachas These South American animals resemble a rabbit with a squirrel tail. They have short, soft fur that is used commercially. There are six species, of which the Chilean or coastal chinchilla (*Chinchilla laniger*) is the most common. Some species and varieties live in the mountains, and these produce a more valuable fur. As a rule, the fur of the chinchillas is in more demand than that of the viscachas.

Both of these animals have been eaten in South America since the time of the Incas, and they have been hunted extensively for food and fur. Some species are now endangered.
(See also RODENTS.)

chiru (*Pantholops hodgsoni*) Sometimes called the Tibetan antelope, the chiru lives on the upland plateau of Tibet and the northern tip of India. The animal prefers an altitude ranging from 12,000 to 18,000 feet. It is surprisingly swift, and can outrun even the wolf. Adult males weigh up to 120 pounds.

The chiru is hunted to some extent, and its meat gets high marks from those people who eat it. But the Tibetans regard the chiru as a sacred animal, and the meat is not eaten by the lamas. The blood of the chiru is believed to have medicinal value, and some people believe that the future can be foretold by the knobs on the chiru's black horns.

(See also ANTELOPE.)

chitons These are armor-plated mollusks that can roll up like an armadillo. About 500 species live in the world's seas, but the largest kinds are found off the coast of Australia and the Pacific coast of the United States. The largest, the giant Pacific chiton *Amicula stelleri*, grows to a length of 8 to 12 inches. It was relished by the American Indians along the Pacific coast. Although it is sometimes called the coat-of-arms snail, it is really not a snail but a separate class of mollusk called Polyplacophora.

cicadas (family Cicadidae, order Hemiptera) These singing insects, comprising over 2,000 species around the world, are commonly called locusts. (True locusts are grasshoppers and belong to the order Orthoptera.) The cicada does, however, form locustlike swarms from time to time. Most of the cicadas are tropical, but the best known of the species is the 17-year cicada of North America and its cousin, the 13-year cicada of the southern United States. At the start of the 17-

year cycle, the female cicada lays eggs on twigs of a host tree. When the eggs hatch, the nymphs drop to the ground and dig in. They live on juices that they suck from the roots of the tree. After 17 years (or whatever their period might be, depending on the species), the grown nymph crawls out of the ground and fastens itself to a tree trunk or some other suitable support. After much labor, the adult cicada bursts the skin and crawls out, ready to fly. It usually emerges at night.

Partly because of the life cycle and partly because of their song, cicadas have figured into folklore and symbolism all over the world. In China and other lands, where the cicada's song is appreciated, the males are even kept in a cage inside the house, much like a songbird or a cricket. (The females don't sing.) The cicada has been used as a source of food, and its value was duly recorded by the ancient Greeks. In modern Cambodia, skilled cicada hunters sit around a campfire at night and clap their hands in unison, making a sound that evidently attracts female cicadas in large numbers, which are caught and eaten on the spot. The cicada can be eaten raw, or it can be fried or roasted. They are often impaled on the point of a stick and roasted over a fire like hot dogs.

(See also INSECTS.)

cichlids (family Cichlidae) These perchlike fish comprise several hundred species, most of which are native to Africa and South America. Because many of them are small and beautifully colored, they are popular as aquarium fish. Some grow large enough to be used for food, and are sometimes raised in ponds.

The largest cichlid, a species of tilapia, is believed to have been raised in ponds by the Egyptians before 2000 B.C. In any case, the largest of the tilapia (*Tilapia mossambica*) grows up to 18 inches long and is an important source of food in some tropical areas, where it is sometimes raised in flooded rice fields. The combination of rice fields and fish culture is believed to have enormous potential as a souce of food for mankind. (See also FISH; NILE PERCH.)

civets (family Viverridae) Some 35 species of the civet live in Africa, Iberia, and South Asia. Some civets feed on birds, rodents, crayfish, and other animals; others eat mostly fruit. Some civets prefer to stay in trees, while others like to forage on the ground; some even swim in the water. Civets are notorious thieves of cultivated bananas, and they also raid corn fields and chicken yards. One species of palm civet is called the "toddy cat" because of its fondness for fermented sap that the natives collect from the palm tree in bamboo tubes. Although they are often called civet cats, and look like cats, they are really members of the mongoose family.

The civet's name is derived from the Arabic word *zabad*, for an unctuous fluid. The animal's secretion, called civet oil in commerce, is important in the manufacture of perfumes; for this purpose, the oil was imported from Africa by King Solomon. Because the production of the oil has been a rather important local industry in some native villages in Africa and in the Orient, the animals are sometimes raised in pens to ensure a steady supply. Civet oil has also been used as an aphrodisiac and in medicine for skin disorders.

As food, the civet has not gained a wide reputation, but it is a major ingredient in the Chinese (Cantonese) stew called the Dragon and the Tiger. Of course, the civet represents the tiger. The snake is the dragon.

clams This large group of two-shelled mollusks move about less freely than scallops, and they don't swim in open water. Nor do they fix themselves to pilings and other structures, as oysters do. Instead, they move along with the aid of a "foot" that extends out of the shell. Although this foot can be used for moving along the bottom, the clam usually digs in the mud or sand and really doesn't stir about much. The foot is usually eaten along with the rest of the clam, but some people do remove them along with the "neck." (One large clam on the west coast of North America, the geoduck, *Panope generosa*, has an edible neck that extends to about 36 inches.)

Most of the clams marketed in the United States are classified as littlenecks, cherrystones, and chowders. The chowder clam is the largest and is usually chopped up for use in chowders and fritters. The littleneck clam is the smallest, and is highly prized for eating on the half shell, either raw or steamed. The cherrystone is of middling size, and is usually eaten steamed.

Apart from size, there are different species of clams. Most of those marketed on the Atlantic Coast are quahogs (*Mercenaria mercenaria*). On the Pacific Coast, razor clams, geoduck clams, Pismo clams, and others are eaten. As on the east coast, the larger Pacific clams are normally eaten in chowder and the smaller ones, the littlenecks or cherrystones, are eaten on the half shell, either raw or steamed.

Actually, there are hundreds of species of clams in American waters, and the ones that fill the markets are merely those that are more plentiful or those that are more easily captured. Probably any American clam taken from unpolluted water is edible and palatable. Anyone who catches their own clams on either the Atlantic or Pacific coasts, or on the Gulf of Mexico, should be sure to check with the local officials about seasons, regulations, and safe clamming waters. People who dig their own clams can often get excellent species that are not available on the market, such as the highly prized horse clams on the Pacific Coast. Other species are not large enough to market at a profit, but are nonetheless very tasty. For example, the tiny rainbow-hued coquina (also called copper belly, butterfly shell, and calico clam) are enjoyed in soup by knowledgeable beach foragers in the southeastern states, especially Florida.

Live clams should be kept on ice or under refrigeration. Any clams that have open shells, or that do not close their shells when they are touched, are probably dead and should be discarded. A high percentage of gaping clams might indicate that the entire batch is old or has not been properly handled. The quahogs will keep for weeks if they are properly cared for and refrigerated, but some other species are much harder to keep in good condition.

Clams are often canned for the market, and some are smoked. Razor clams are common in British and European coastal areas, but the British and Europeans much prefer the cockles, or heart-shaped clams, which have been important sources of food in the past. Clams are eaten in Japan, and a clam soup called *hamaguri* is served at wedding banquets; the idea is that the two halves of the clam close together, making a whole that is better than the sum of the two parts. Clams were also eaten by ancient man over most coastal regions of the world, as has been made clear by piles of shells at numerous kitchen middens.

(See also BIVALVED MOLLUSKS.)

climbing perch (*Anabas testudineus*) These small fish, growing up to about 8 inches long, have an organ that enables them to breath air. They live in both fresh and brackish waters of Southeast Asia. They can push themselves over the ground with their almost footlike pectoral fins, and can even climb rocks and stumps, as their common name implies. During rainy seasons, the climbing perch come out of the water to feed on worms in fields and gardens.

The climbing perch is caught for food and, in dry weather, many of them can sometimes by found in mudholes left as streams dry up and ponds evaporate. They will even bury themselves down in the mud; in extreme dry spells they will become dormant. In India they are dug out with scoops.

(See also FISH.)

cockles These are bivalved mollusks with a heart-shaped shell. Several species exist, and some other types of clams are called cockles. The most important one for eating is *Cardium edule*, which is plentiful and commonly eaten around the British Isles. It also grows along the Atlantic coast of France, usually at the mouth of rivers, as well as in the Mediterranean. Other edible species of cockles live along both the Atlantic and the Pacific coasts of North America. Most of the cockles that have been eaten on a regular basis in the past live in very shallow water near the shore. Modern scuba diving equipment has led the way to foraging for other species that live in deeper water and are often larger. All of the true cockles belong to the Cardiidae family, which derives its name from the Greek *kardia*, meaning heart.

Although a number of cockles are eaten, *Cardium edule* has been the most important cockle for the European table for a long time. These are small cockles, about 1 inch long on average, which are often steamed before they are opened. They are eaten raw and also used in soups or stews. Many are canned. Not long ago, cockles were called the poor man's oysters and were an important part of the coastal cuisine of the British Isles. These days, they are more expensive and are considered to be appetizers. A Japanese cockle called *torgai* is of commercial importance. Primitive man also ate cockles, using the larger shells for utensils.

cockroaches (family Blattidae) Cockroaches of one species or another have hidden out in dark places of the earth for some 250 million years. They are, in fact, the oldest of the fossil insects. There are some 1,600 species in the world today, including some with a wingspan of more than 5 inches and one that actually hisses.

During its long history, the cockroach has been used as food, either on a regular basis or as survival fare. Such cases have not been widely documented, but the use of the cockroach for remedies for various human ills is common. U.S. journalist and essayist Lafcadio Hearn, for instance, reported in 1886 that some people in Louisiana fried cockroaches in oil and garlic, then ate them to alleviate indigestion. Cockroaches were also used as food by the native Australians and South Americans. Today they are reported to be eaten by gourmets in China. (See also INSECTS.)

coelacanths (*Latimeria chalumnae*) Sometimes called a living fossil, the coelacanth belongs to a group of fish that were believed to have been extinct until they were redisovered by scientists in 1938. At that time, the good boat *Nerine* docked in South Africa to unload a catch of fish, and the curator at the East London Museum came looking for specimens. What she found was a 5-foot, 100-pound fish with a sort of tripple tail and odd, meaty, flipperlike fins. Since it had a large (but degenerate) lung, it was known to be kin to the lungfish; but clearly it was a separate species. Further investigation fixed it in the order Coelacanthiformes, believed to have been extinct for 60 million years.

Nevertheless, the fish has been caught and eaten in recent years by the Comoro islanders offshore from the Malagasy Republic of Africa, where the fish is called *kombessa*. This living fossil is usually eaten either dried or salted instead of fresh.
(See also FISH)

conchs Pronounced "conks," these large marine snails with beautiful shells are really a stable food of the West Indies. The extremely tough meat of the conch must be tenderized, usually by pounding, before it can be chewed easily. Some modern cooks put the conch into a pressure cooker until it is tender. Others put it into a washing machine to be batted about by the paddles. Conch salad is very popular in the Bahamas, where, according to Jessica B. Harris's *Sky Juice and Flying Fish*, the mucilaginous string that attaches the conch to its shell is considered by the natives to be an aphrodisiac and is sold at conch stands. At one time the conch shell was widely used as a horn, but these days the shells are sold mostly to tourists.

Usually, it is the large queen conch (*Strombus gigas*), which grows up to 15 inches in length, that is eaten in the Caribbean. The larger horse conch, which grows in Florida, is edible, but has a bitter flavor. The helmet conch also has a bitter flavor, but it can be eaten and in the Mediterranean its shell is used to make cameos. Along the eastern coast of the United States, the whelk—a similar marine snail—is also eaten and is often called a conk, especially along the Outer Banks of North Carolina, where there is an old saying: "If you ever eat a conch chowder in Carteret County, you'll never want to leave."

conger eels (*Conger oceanicus*) Female congers grow up to 8 feet long, whereas as a rule the males reach only 2 or 3 feet. (The females of other eel species are also larger.) Female congers also outnumber males. Congers tend to live in the warm and temperate zones of the Atlantic, Pacific, and Indian oceans. They have sharp teeth and are definitely predators, much preferring to eat fresh fish, squid, crabs, and octopuses. They feed mostly at night and hide during the day in rocks, wrecked ships, and other suitable cover.

Because of their habits, congers are not taken in large numbers by commercial fishermen, but some are caught on long lines and in trawl nets. The fish is sometimes taken by sport fishermen. The anglers of the British Isles—around which the conger seems to thrive in spite of occasional cold-water kills in the North Sea—are fond of tangling with the big eels. Large specimens are, of course, difficult to land, and they have a way of wrapping their tails around any object in the water. Once the conger is brought to the surface, it spins rapidly, often twisting itself right off the hook. When boated, the thing fights and snaps its jaws and is difficult to kill. In short, the conger, like the moray, is dangerous both in the water and out.

Conger steaks are sometimes sold in fish markets, and some anglers catch their own congers for table fare. The meat of the conger is usually recommended for use as an ingredient for stews and soups, such as the bouillabaisse.

(See also EELS; ELECTRIC EELS.)

coots A 2-pound bird of the rail clan (family Rallidae, genus *Fulica*), the coot sits and bobs on the water like a wild duck, but it has a beak like a chicken. The bird is, in fact, sometimes called a mudhen or moorhen. The coot feeds mostly in the water, diving if necessary, on bits of vegetation, seeds, mollusks, and so on. As a rule, the bird prefers shallow water over a mud bottom, and is especially fond of shallow lakes that are lined with reeds or other vegetation in which they can hide. They take to brackish water bays and inlets as well as fresh waters.

The bird's feet aren't webbed, but they have wide toes with flaps that aid in swimming and walking on muck or floating vegetation. Before gaining flight, the coot makes an awkward, noisy run across the top of the water. Once airborne, the coot is a strong flier. Ten species grow around the world. The common American coot (*Fulica americana*) ranges from central Canada to South America. Another species ranges from Ireland, across Europe, and on to Japan.

Coots are sometimes taken by sportsmen, and some states have very liberal game laws and large bag limits on the bird. But, alas, the coot is not highly regarded as a game bird or as table fare in most places, and certainly not in America. Yet, the French culinary tome, *Larousse Gastronomique*, gives the bird high marks, saying that the meat—although dark and dry—is mild and very good when eaten fresh, preferably after cooking with a little bacon to add some moisture. At one time, the bird was sold in European markets, and was classified as Lenten fare. The key to understanding the coot's mixed rating as food is the fact that the skin, not necessarily the meat, has a strong flavor. For the best culinary results, hunters should skin and draw the bird as soon as it is shot. Knowledgeable coot lovers will not throw away the innards. The bird has a very large gizzard that is good to eat, and the liver is also tasty.

(See also BIRDS.)

Coot Gizzard Pilau

In *Cross Creek Cookery*, Marjorie Kinnan Rawlings set forth the following recipe: "A coot liver and gizard pilau is made simply by cooking available coot livers and gizzards with enough rice to feed as many people as need feeding!"

---

cormorants (*Phalacrocorax carbo*) A large seabird of worldwide distribution, the cormorant is a notorious fish eater, able to be trained to catch fish and bring them to a master. Before going fishing, however, a band must be put around the bird's neck to keep it from swallowing the fish. The sport of fishing with the birds was once popular in Europe, and the Master of the Cormorants was an official position in the royal household. Cormorant fishing is still practiced to a very limited extent in Japan, China, and other parts of the Orient. Often, the best fishing takes place at night, since torches are used to attract fish to within easy swimming range of the cormorants.

In addition to providing fish for the table, the cormorant is edible when young. (The mature bird, however, is said to be unedible, although a few have no doubt been consumed in dire situations.) The young birds, or squabs, as well as the eggs, are taken for food at the rookeries, especially among the seacoast and island peoples of northern Europe. Such practice, however, might be frowned upon in Peru, where the birds are fully protected by law because they provide high-grade commercial guano on the islands off the coast.

(See also BIRDS.)

crabs Many species of crabs live in all the seas as well as in many large rivers and lakes. Some even live on land and climb trees in search of food. A wide number of crabs are eaten here and there around the world, but only a few of the over 4,500 species are important commercially. In Europe, the edible crab (*Cancer pagurus*) is taken in large numbers in pots or traps, and the British alone catch 6,000 tons annually. On the east coast of North America, the blue crab (*Callinectes sapidus*) is taken in pots, by handline, and by trawl nets. (It is a swimming species and can therefore be netted.) The soft-shelled blue crab is highly regarded as table fare and is often marketed to restaurants; hard-shell blue crabs are sold fresh, frozen, and canned.

A number of other crabs are important commercially, including the dungeness crab, the "king crab," the "snow crab," and the stone crab. Local favorites are defended in many coastal regions. Some swimming crabs, closely related to the blue crab of the Atlantic, can be taken in nets and are important sources of seafood in the Indo-Pacific region. In the North Pacific, the large king crabs are taken by trawl nets and are frozen at sea on factory ships.

(See also LAND CRABS; PEA CRABS.)

crayfish Crayfish live in the freshwaters of all continents except Africa, and they are highly esteemed as food in several parts of the world. Their meat compares well with lobster and, of course, they look like miniature lobsters. However, since the crayfish are usually much smaller than lobsters, only the tails are eaten.

Commercially, crayfish are quite important in France and in parts of the United States. They are farmed in ponds and are also caught commercially from the wild with the aid of traps. Over 300 species of crayfish live around the world, but many of them are too small to be of much food value. The largest, the Tasmanian crayfish (*Astacopis gouldi*), attains a length of 16 inches and a weight of 8 pounds. Often, several kinds live in the same region, especially in the United States. Louisiana, for example, has 29

species, but the most important commercially is the red swamp crayfish of the southern part of the state. The red swamp crayfish is often raised as a second crop in flooded rice fields. There is also a good deal of crayfish farming in the northwestern states of Washington and Oregon.

The crayfish is honored in several annual feeds and festivals. In Breaux Bridge, Louisiana—billed as the crayfish capital of the world—the jubilee features such events as crayfish races and crayfish-eating contests. In Scandinavia, an annual banquet honors the crayfish.

The crayfish is often spelled crawfish, and is called crawdad, mudbug, yabbies, creekcrabs, and so on. In Australia, a popular river species is called Murray River Lobster. Before refrigeration and modern canning methods were developed, crayfish were a very important source of "seafood" for people who lived a long way from the sea. They were especially popular in medieval Europe.

Some crayfish burrow into the ground, forming a chimney at the entrance to their hole. Sometimes these chimneys are a foot high. In some areas, the burrowing crayfish are destructive to crops.

None of the true crayfish lives in salt water, although the spiny lobster is often called a crayfish.

(See also LOBSTERS.)

crickets About 1,500 species make up the crickets, family Gryllidae. These range in length from ⅛ inch up to 2 inches, and all have large hind legs designed for jumping. Most of the crickets chirp, and they are very often heard in houses, barns, and other structures built by man. In fact, crickets are kept as pets in Asia, especially Japan, as well as other places where they are valued for their song. House crickets seem to have a fondness for dry, warm places and often frequent the hearth. In many areas, the house cricket is believed to bring good luck, and therefore killing one is thought to bring bad luck. In Malay, the cricket is believed to be the spirit of the dead.

In Japan, as well as in China and some European countries, crickets are kept as pets in special cages made of bamboo, coconut shell, and other materials. The more expensive cages often have lids of carved ivory. Some were made of gourds that were grown inside bottles in order to shape them.

In China, cricket fighting is a time-honored sport that goes back at least a thousand years. Two fierce crickets, often specially bred and trained, are pitted in a ring for a fight to the death. Winners are prized for breeding purposes, much like thoroughbred horses, and official records are maintained. Several days before the fight, the trainer allows mosquitoes to bite him on the arm. The mosquitoes are caught and fed to the crickets in the belief that the trainer's strength will be transferred to the cricket. In addition to mosquitoes, the crickets are fed a special diet of rice, fresh cucumbers, boiled chestnuts, lettuce, and various seeds. The crickets are carefully weighed before each fight and classed as heavyweight, middleweight, or lightweight. In the pit, the trainers hold the crickets back and infuriate them by tickling them with a special tickler made by inserting rat hairs in ivory or bamboo handles. At one time, cricket fighting was a very popular pastime in China, and large sums of money were wagered. In a great championship fight, a cricket named Genghis Khan from Canton won $90,000 in a single bout.

Although true crickets don't often swarm in clouds like the locust or the grasshopper and therefore are not easily captured in large number, they are nonetheless eaten in some places. At one time in Burma, a brown cricket called *payit* was captured, fried and sold in markets. (When fried they were called *payit-kyaw*, meaning fried cricket.) According to *Insect Fact and Folklore*, they were purchased by the wealthy Burmese to feed wandering Buddhist monks. According to Waverley Root, the foods writer, the best crickets are found in Thailand and the state of Oaxaca in southern Mexico.

(See also GRASSHOPPERS; INSECTS.)

crocodiles See ALLIGATORS AND CROCODILES.

crustaceans See BARNACLES; CRABS; LOBSTERS; PEA CRABS; SHRIMPS AND PRAWNS; STONE CRABS.

cuttlefish (*Sepia officinalis*) Although these cephalopod mollusks are almost entirely missing from American menus and even cookbooks on seafood, the common European cuttlefish is highly regarded as table fare in the Mediterranean and in other parts of Europe. This species grows to about 2 feet in length and somewhat resembles an octopus. Like the octopus, it emits a black ink as an escape mechanism. This ink is sometimes used as an ingredient in cuttlefish recipes. Similar species—about 100 all together—grow in the tropical and temperate seas of Asia. The waters around Japan have a number of species. In China, cuttlefish are very important commercially. They also make up a significant fishery in the Mediterranean, and are marketed fresh or canned. Cuttlefish breed in deep water, but most of their lives are spent in shallow waters of coastal areas. They are often caught by sport fisherman, who use lights to attract them to the surface. During the breeding season, some fishermen slowly troll a female behind the boat to attract males.

The cuttlebone of commerce also comes from this fish, and is in fact an internal shell. It has been used in medicines, and the ladies of ancient Rome used powdered cuttlebone as a cosmetic. In recent times, the cuttlebone has been marketed in pet shops as a source of calcium for pet birds.

The cuttlefish receives high culinary marks, but it must be prepared properly. Usually, the meat is tough and must be pounded or otherwise tenderized.

(See also OCTOPUSES AND SQUID.)

D

deer One or another of thirty-six species of deer live on most of the world's land masses except for Africa, Arabia, and the eastern part of Greenland. All of the deer make good eating if they are handled properly. In Europe and parts of Asia, the relatively small roe deer (*Capreolus capreolus*) gets high culinary marks as table fare, although the much larger red deer (*Cervus elaphus*) is more highly regarded for the chase.

In America, the white-tailed deer (*Odocoileus virginianus*) is very popular among both hunters and meat eaters. At one time, the whitetail disappeared from a good part of its range, due to overhunting and other problems. Thanks to modern game management techniques, it has been restored to most of its former range. In fact, it is now considered to be a pest in many agricultural areas and in some suburbs, where it eats gardens and potted plants and poses a hazard to automobile traffic.

Some of the deer have been domesticated, while many others are "managed" to such an extent in parks and preserves that they may well be somewhere between wild and domesticated. Also, many farmers in America and other parts of the world now consider the resident deer to be a crop to be managed and sold, either in the form of meat or in hunting rights. In fact, virtually all of the various state game and fish agencies, as well as those of federal governments and large landholding companies, speak of harvesting deer as if they were a crop. The same thing has happened in Europe, where fallow deer and others are indeed harvested for profit.

In any case, the meat of all deer is called venison. It is good eating if it has been wisely selected and properly handled prior to and during cooking. Unlike modern beef, venison has little fat marble in the tissue.

The smallest deer, the southern pudu (*Pudu pudu*), an endangered species that lives in the foothills of the Andes, weighs only about 17 pounds. The largest of the deer, the moose (*Alces alces*)—which is called elk in Europe—weighs in at 1,750 pounds. The moose lives in the northern parts of North America and northern Europe, as well as in parts of Siberia, Mongolia, and Manchuria. The moose has also been introduced to New Zealand. In fact, other species of deer, including the whitetail, have also been introduced in New Zealand and Australia, where none existed before.

Many cave paintings of deer have been discovered, some dating back 14,000 years. Venison is also believed to have been very important in the diet of primitive man.

(See also MUSK DEER; REINDEER; WAPITIS.)

deer mice These graceful animals are neither deer nor mice but, being small ruminants, they are more kin to deer than to rodents. They are chevrotains, comprising four species in central Africa, India, Sri Lanka, and Southeast Asia, including the Philippines. The water chevrotain (*Hyemoschus aquaticus*) is the largest species, growing up to 29 pounds, and lives only in Africa. The lesser mouse deer (the smallest of the four) weighs about 5 pounds and lives in the tropical rain forests and mangrove swamps of Southeast Asia.

All of the chevrotains are eaten throughout their range, but they are secretive animals—active mostly at night—and therefore not primary targets for hunters.

dogs (*Canis familiaris*) Because dogs are the oldest of the domesticated animals, it is somewhat surprising that they are not more generally accepted as table fare. In emergencies, they are eaten by the Eskimo, and throughout history they have been used as food in cities under siege or in times of famine. Some of the tribes in the Philippines and throughout Polynesia have eaten dog on a regular basis in the past. In China dogs of a certain breed—the chow chow—are fattened especially for the table, and this, it has been suggested, is the origin of the American Navy slang term, "chow down." The ancient Romans also fattened dogs for the table, and reportedly were fond of puppies.

In any event, Marco Polo reported a number of incidences of the peoples of the steppes between Europe and China eating dogs. He wrote, for example, of the inhabitants of Kinsai: "They eat all kinds of meat, even dogs and other foul animals which no Christian would dream of touching." The ancient Chinese are known to have eaten both wolves and foxes, and they made a sort of crackling by frying the fat from wolves. They also ate pieces of dog liver barded with the animal's own fat and grilled over charcoal.

The American Indians ate dogs, and the Aztecs apparently raised them for food and sold them at the market in Tenochtitlán. When the Spanish came to Mexico, the dog and the turkey were the only domestic animals used for food. Further south, the Incas hunted fox for food, and the domesticated dogs were widely eaten by the peasants—but not by royalty. The Maoris of New Zealand ate dog as well as rats. Captain James Cook said that the Maoris raised dogs for no other purpose than to eat them. The Maoris also used the hides of dogs, since there were no large leather-bearing animals on the island.

In any case, the dog was one of the earliest domesticated animals, and may have been tamed first as a source of food, as a hunting partner, or for warmth. The latter theory is explained by the modern day Eskimo practice of bringing in one or more dogs on cold nights. Thus, a three-dog night would be colder than a two-dog night.

Dolphin

dolphins (*Coryphaena hipurus* and related species) These fish, not to be confused with sea-going mammals, can weigh up to 70 pounds. They feed around mats of seaweed and other floating objects, and are especially plentiful in the warm North Atlantic waters along the Gulf Stream, as far north as Prince Edward Island. They also live in other tropical and subtropical salt waters around the world. A good fighter and a voracious feeder, the dolphin is popular with sportsmen. It is also highly praised by gourmets. The fish are often marketed under the Hawaiian name, *mahi-mahi*, and are sometimes found in restaurants. Having medium oily, firm flesh, the dolphin is one of the better saltwater fish, provided that it is gutted and iced down shortly after it is taken from the water.

dolphins and porpoises These mammals are really small members of the whale group (mammalian order Cetacea). The porpoises include six species, all of which are edible. The flesh is said to resemble pork, but it is quite oily and has an odor that is not agreeable to some people. According to *Larousse Gastronomique*, the porpoise was once sold in the markets of Paris during Lent, and was prepared in various ways. The meat was also eaten in Iceland and Newfoundland, and is reported to have been quite popular in 15th-century England. The oil of the porpoise has also been used by man, often as fuel for lamps. The oil doesn't gum up badly; it withstands quite low temperatures, and was at one time widely used to lubricate watches.

The porpoise tends to prefer coastal areas of the tropical and temperate seas, whereas the dolphin is often found in deep-sea waters. All of the porpoises (family Phocoenidae) are comparatively small, and the common harbor porpoise (*Phocoena phocoena*) averages from 4 to 6 feet in length. In general, the porpoise looks more or less like a dolphin except that it doesn't have the characteristic beak nose and the hump isn't as pronounced.

Dolphins are also edible, and some authorities rate the meat as excellent. They are much more widely distributed than porpoises, especially in the open seas. The family Delphinidae also includes the killer whale and a few other small whales in addition to the dolphin, making a total of 32 species in the entire family. The five species of river dolphin are in yet another family (Platanistidae). Although all of the dolphins are edible, the main danger to these animals comes from the seines and nets used by commercial fishermen after tuna and other fish.

The dolphins are perhaps the best known of the whales and other marine mammals, owing largely to the popularity of the *Flipper* television show and films, and to the several outdoor aquariums in Florida, California, and other places. The dolphins that are usually used in these establishments are the bottle-nosed dolphins (*Tursiops truncatus*), which always seem to smile. The dolphin was also well known to ancient man, and especially to seafaring folk. Many of the seagoing peoples believe that eating either the dolphin or the porpoise brings bad luck.

(See also WHALES.)

donkeys A number of donkeys and wild asses roam about the world, from Africa to Iran to the mountain peaks of Tibet and the plains of Mongolia. The donkey is really a domesticated wild ass (*Equus asinus*) of northeast African origin. All donkeys are excellent venison and are sometimes hunted for both sport and food. Donkey meat was highly praised by the Romans, and in Italy, donkey and mule are smoked and sold. In France, the donkey has been raised and fattened expressly for food. The milk of the she-ass, said to be quite close to that of humans, was something of a fad there at one time.

The donkey was used as a beast of burden by the Egyptians thousands of years before the Christian era, and it was an important animal in biblical times—but not for food. It was considered to be unclean, and its flesh was forbidden as food. During a great famine in Samaria, however, an ass's head sold for "fourscore pieces of silver" (II Kings, 6:25).

The mule is a cross between a male ass (jackass) and a female horse (mare), and has also been highly praised as table fare. The mule is quite old, and was used as a beast of burden in Asia Minor at least 3,000 years ago. Both the mule and the ass are considered to be better eating than the horse.

(See also HORSES.)

dormice A quiet creature, the dormouse usually stirs about only at night and hibernates during the colder months. Like the squirrel, it haunts bushes and trees and stores nuts for use during hibernation. There are several species of dormice, ranging in size from mouselike to the edible 10-inch European *Glirinae glic* and ranging in habitat from Europe westward across Asia to Japan.

At one time, the dormouse was highly regarded as food in Europe. The meat was considered to be at its prime just before hibernation, when the animal had fattened itself up for the long winter months. The Roman epicures even captured dormice and fattened them in special cages before cooking them with honey and poppy seeds. They were usually served as an appetizer prior to a feast.

(See also RODENTS.)

doves See PIGEONS AND DOVES.

dugongs See MANATEES AND DUGONGS.

duikers Also called duikerbucks, these small African antelope comprise 17 species. The name *duiker* comes from the Dutch and refers to the animal's way of ducking into thick brush and disappearing whenever danger threatens. The common duiker (*Sylvicapra grimmia*) lives on the savannas and open bush country; the others live in the forests. All species live south of the Sahara.

According to *The Encyclopedia of Mammals*, the main threat to the duikers is subsistence hunting: "Duiker meat is well-liked by humans and duikers are easily dazzled at night with lights, making them comparatively easy to shoot or capture. Some species are also captured by driving them into nets."

(See also ANTELOPE.)

E

echidnas Sometimes called spiny ant-eaters, these strange creatures come in two kinds: the long-beaked echidna, *Zaglossus bruijni*—weighing up to 22 pounds—and the short-beaked echidna, *Tachyglossus aculeatus*—weighing up to 17 pounds. The longbeaked species live only in the mountains of New Guinea. The shortbeaked kind live in Australia and Tasmania, as well as in New Guinea. Designed for eating ants and other small insects, the short-beaked echidnas have long tongues that are coated with a sticky secretion. The long-beaked echidnas, however, have horny spines in a groove in their tongue; the spines are used to hook worms, which make up most of the animal's diet.

Like porcupines and hedgehogs, the furry echidnas use their heavy spines as a main line of defense. When threatened, they either ball up or wiggle themselves straight down into soft dirt, leaving the spines sticking up above the ground level. Also, echidnas can wedge themselves in hollow logs or crevices by erecting their spines.

Both kinds of echidnas lay eggs and, along with the platypus, are the only mammals that use this form of reproduction. With the echidnas, the egg is "laid" into a pouch, where it hatches within about 10 days. The long-beaked echidna is not very plentiful, but the short-beaked kind is common in most of its range, although it is nocturnal and seldom seen. Both kinds of echidnas are edible and are indeed hunted for food.

eels The common American eel, *Anguilla rostrata*, is not as appreciated as the European eel, *Anguilla vulgaris*, which is widely eaten and highly praised in England, France, and other countries. Yet, it is almost the same exact fish. Both kinds are born in or near the Sargasso Sea in the Atlantic ocean southeast of Bermuda.

After the eggs hatch, the American eels—tiny larval creatures, so thin and transparent that newsprint can be read through their bodies—begin a year-long journey to the coast of North America. Then large numbers of the eels—elvers now, measuring from 2 to 3 inches long—enter the brackish water marshes and estuaries from the Gulf of Mexico to the Gulf of Saint Lawrence. Apparently, the male eel stays in the brackish waters, but the larger female swims on upstream, climbing waterfalls, dams, and rocks. Some end up in small creeks a thousand miles from the sea, and sometimes they are found even in ponds. The female eels stay in fresh water for several years, or at least until they mature. Then they migrate downstream into the brackish waters—where the males have been waiting—and head into the open sea. Somehow, the adult eels find their way back to the spawning area, where the process is started again. Apparently the adults die after spawning.

The European eel does pretty much the same thing, but its larvae head for the coast of Europe, making an even longer journey. After hatching, the larval eels drift with the Gulf Stream until they reach the Atlantic coast of Europe, a 5,000 mile journey that may take up to three long years.

After maturing in the rivers, the European eels head back downstream and toward the spawning area in the South Atlantic. While still in the rivers, they are called silver eels. Silver eels are considered to be the best eating because they fatten up before undertaking the long ocean journey.

In addition to the American and European eels, a number of other species grow around the world, some of which live only in salt water. In some areas, eels are highly prized, and they are often kept alive until moments before they are cooked. Although some ethnic groups in New York and other cities eat eels in America (often before Christmas), the fish is grossly underutilized in most areas. Maybe its snakelike appearance and slimy skin turn people off. Perhaps some of its unpopularity may be due to the fact that Americans are fond of frying fish and this is the worst way to cook eel because it is very, very oily.

Eels are best when they are smoked or grilled. They are also widely used in stews. In Provence, France, baked eel—cooked with wine, chopped leeks and spices—is traditionally eaten on Christmas Eve. In Pisa, Italy, a specialty called *cieche alla pisana* is made by cooking tiny eels in oil with sage and garlic. The eels, or elvers, are taken during migration at the mouth of the Arno River. In Denmark, eel parties are said to be popular occasions; the Danes will drive for miles to inns that have a reputation for serving up good eels, which are usually eaten with potatoes.

Clearly, Americans are missing out on a good thing by not making better use of the large population of eels on the Atlantic side of the country. This may change. In Florida, an eel farm raises and fattens elvers in ponds and markets them—2mostly to ethnic restaurants—where they command high prices.

Although they are not true eels, lampreys are also eaten and are highly prized as food in England and some other countries. They are not consumed in the United States, however, although they literally infested the Great Lakes a few years ago and were a great problem to the sport fisherman. Even the eggs of lampreys were highly praised by Roman epicures.

Some large aquatic salamanders have only tiny feet and are sometimes eaten as lamper eels or conger eels.

(See also CONGER EELS; ELECTRIC EELS; MORAYS.)

eggs To modern man over much of the world, the eggs that are conveniently available in supermarkets are assumed to be from the chicken. This has not always been the case. Duck eggs, turkey eggs, guinea eggs, peacock eggs, and others have been raised on the farm and sold in markets around the world. Other eggs are also eaten here and there, such as quail eggs in China and pigeon eggs in France.

Not many years ago, the eggs of wild birds were widely hunted and marketed in untold thousands. Egging was especially productive on offshore islands, where many seabirds nest because they are more or less protected from foxes and other predators. Some of these bird eggs were marketed and eaten as such, but others were used in different ways. During the 19th century, for example, eggs of the sooty tern were taken from Dry Tortugas and sold by the gallon to commercial bakers.

Wild bird eggs are not only still eaten in some countries, they are highly re-

garded in the cuisines of the Netherlands and parts of Scandinavia. Of
course, bird eggs have been an important food to the Eskimo, and the ostrich
egg is eaten in Africa.

Turtle eggs have also been sold commercially in the past; they may still be
available in some parts of the world,
legally or illegally. These are usually
obtained by raiding the nests of large sea
turtles on sand beaches. Many fishermen and people who catch (or trap)
freshwater turtles for their own use also
eat the undeveloped eggs. Often, farm
wives cook undeveloped chicken eggs and
put them into giblet gravy. Other reptile
eggs are also edible and are highly regarded by the Arabs and Indians. Alligator
and crocodile eggs are eaten by the natives
throughout their natural range, although
such practice may be illegal in some
areas.

From the sea we have fish roe, including the highly prized caviar. In the
American northeast, shad roe are cooked
in a number of ways. Salted mullet roe
were eaten by the ancient Egyptians as
well as by modern epicures. A pike-
shaped fish in Africa, *Hepsetus odoe*, deposits its eggs in a floating foam
nest—which is considered to be a great
delicacy. Roe of bonefish, tarpon, and
many other fish are also eaten. Even lam-
prey eggs are eaten.

Crab eggs are the distinguising ingredient in she-crab soup. In China, yellow crab eggs are used in recipes for
asparagus and Chinese cabbage. Lobster eggs are an added bonus for many
diners. The eggs of sea urchins are prized
all around the world, and even
waterboatman eggs are marketed in
Mexico. Ants' eggs are eaten in Laos and
snails' eggs are served in Paris.

In any case, most eggs make tasty eating and are highly nutritious, except for
people who need to watch their cholesterol levels. But not every sort of egg is
safe to eat. The roe of garfish and
cabezones are highly poisonous. All
birds' eggs are edible if they are fresh,
and in fact can be eaten at any stage of
embryonic development. In recent years,
however, there have been some deaths
reported from salmonella poisoning resulting from eating commercially raised
chicken eggs from the supermarket.
Usually, the salmonella comes from
cracked eggs that are eaten raw or only
partly cooked.

eiders Several species of these large seabirds live around the fringes of the Arctic. The common eider (*Somateria
mollissima*) grows 2 feet long and inhabits
northern Europe and Iceland. Other
species also live in the same region, as
well as around the North Atlantic to
Greenland. One species, the American
eider, lives as far south as Maine, where
it is taken by wing shooters. In the Pacific, similar eiders live in Alaska, Siberia,
and the Aleutian Islands. One large species, the king eider, is truly circumpolar.

All of these birds are used as food
throughout their range, and some peoples, such as the Eskimo of Greenland,
depended upon them as a part of their
standard fare. In fact, eiders pickled in
a blubber-lined sealskin and frozen in
a cache for several months were considered to be prime eating. The eggs (also
preserved for future use) were just as
important as the meat, and perhaps
more so in some areas.

The down of eider has been used by
native peoples to make warm clothes and
bedding, and the material has been used

commercially for jackets, sleeping bags, and so on. Actually, the eiderdown is harvested from the birds' nests, which are built on the ground and lined with fine down feathers plucked from the hens' breasts. The down keeps the eggs warm during the hens' absence. If an eider nest is robbed, both the eggs and the down will usually be replaced at least once. Gathering the down is a fairly important industry in Iceland, where the practice is regulated by law to ensure a continuing resource. Eiderdown is also gathered commercially in some other remote arctic regions, such as the Novaya Zemlya.

(See also BIRDS.)

elands These are the largest of the antelope. The common eland (*Taurotragus oryx*) reaches a height of 6 feet at the shoulders and weighs up to 1,200 pounds. It once roamed the grasslands over virtually all of Africa, but now it is found mostly in preserves and ranches. It is excellent eating and has been more or less domesticated at one time or another as a draft animal, as a source of meat, and as a source of milk and dairy products.

The giant eland (*Taurotragus derbianus*) is even larger than the common eland and the males may weigh almost a ton. It is more of a woodland species and isn't a range animal. Both sexes of both species have long spiral horns.

(See also ANTELOPE.)

electric eels (*Electrophorus electricus*) These eellike fish (which aren't true eels) live in the freshwater streams of the northern half of South America. They prefer sluggish, well-shaded pools. The electric eel can grow to a length of 9 feet, but 3-footers are average. The fish generates an alternating current in its body, which passes from its tail (positive) to its head (negative). This jolt has been measured, for brief periods, at up to 500 volts—enough to stun a man. Nevertheless, the fish is sometimes used as food by the natives.

(See also CONGER EELS; EELS; MORAYS.)

elephant fish This group comprises over a hundred species of fish with elephantlike snouts, and all are native only to the fresh waters of Africa. The snout, with a small mouth, is inserted into the mud and into holes among rocks while the fish searches for food. Some elephant fish grow to only 6 inches in length, while others get considerably larger. The fish has the largest brain of the lower invertebrate animals, and it can generate a mild electrical field that acts somewhat like a radar navigational system. One species, *Mormyrops delicious*, is listed in *McClane's New Standard Fishing Encyclopedia* as growing up to 5 feet in length. The last part of the scientific name is difficult to explain because its flesh is said to be mushy and tasteless. The meat of the larger fish is usually eaten dried. Some women of East Africa believe that eating elephant fish will cause their children to be born with snouts.

(See also FISH.)

elephants In modern times, both the Asian elephant (*Elephas maximus*) and the African elephant (*Loxodonta africana*) have been hunted for sport and ivory as well as for food. Both species have been used as beasts of burden, dating back 5,000 years to the Indus Valley. Hannibal used African elephants in his war with the Romans, and as late as World War II elephants were used to haul heavy military equipment. The smaller Asian elephant has been used more extensively as a beast of burden, a fact based in part on the skill of the mahouts. The elephant hasn't been used often in Africa, although the practice was initiated by the French and the Belgians in the Congo. The African elephant has much larger tusks (from 6 to 8 feet in length as compared to 4 or 5 feet for Asian elephants) and has been hunted extensively for its ivory. Although illegal in most areas, ivory hunting (or poaching) was made even more lucrative by modern weapons. Both species, of course, have been threatened by the loss of habitat, although a number of parks and wildlife preserves have been established in both Africa and Asia.

In Africa, the elephant herds within the national parks have overpopulated and have been "thinned" out or culled from time to time. When this has happened, the state sold the valuable ivory at auction and the meat was dried and used locally. The trunks and the feet are considered to be delicacies. Elephant fat was made into cooking oil, and the hide was used in leather goods.

Elephant meat has been consumed by mankind for a very long time. Primitive man hunted elephants (and mammoths) for food. In recent times, the Akoa Pygmies of Gabon hunted elephant with the aid of poison arrows. The elephant hunt was a way of life, complete with ritual and song. The Pygmies enjoyed the excitement and danger of the kill, as well as the feast that followed success. The women and children followed the hunters and, when an elephant was felled, the women and children joined in a ceremony around the carcass. Then the feast began.

First, the trunk was removed and smoked over a fire pit. Some of the meat was smoked for later use, but most of it was eaten where it fell. After all, a large elephant will weigh up to 13,000 pounds. (The largest on record, killed on the Cuando River district of Angola, weighed 12 tons.) Of course, the carcass drew flies, and sometimes the meat was infested with maggots before the feast was even finished. The natives ate the maggots along with the elephant meat.

The Mbuti Pygmies of Africa also hunted elephants for food, using both bows and heavy nets, but without the ritual of the Akoa Pygmies. The Mbuti hunters and their families took along kettles to boil the meat. To speed things up, they sometimes pulled out the innards and then climbed into the elephant's body cavity in order to cut off chunks of meat for the pot. They ate all that they could hold, then dried the rest of the boiled meat and took it back to the village.

Generally, the trunk is considered to be the best part of the elephant; it has been compared favorably with ox tongue.

Elephant-shrew

elephant-shrews These small animals look like long-legged shrews with elephantlike snouts and long ratlike tails. Several species live in the southernmost third of Africa and along the coastal area of North Africa. The largest species, the golden-rumped elephant-shrew (*Rhynchocyon chrysopygus*), is about a foot long and weighs a pound. All of the elephant shrews are terrestrial and feed on invertebrates.

Since the animals are small and seldom seen by man, they would seem to be an unlikely source of food. But they have a habit of using trails through their usual feeding area for normal coming and going and, especially, for fast getaway when they are threatened by other animals. The trails are well defined, making it easier for trappers to catch the elephant-shrews in simple snares for use as food. They can be skinned, gutted, and cooked like squirrels and other small animals.

emus (*Dromiceius novaehollandiae*) In size, this flightless bird stands second only to the ostrich. Several subspecies of the emu once lived in Tasmania and the Kangaroo Islands but are now extinct. Like the ostrich and the South American rhea, the emu has been actively pursued for its flesh and eggs and is a bird of the open plains. It feeds on roots, fruits, and leaves; in modern times it has developed a taste for wheat, on which it feeds in the fields of western Australia.

The aborigines stalked the emu and threw bone-tipped spears or boomerangs at it. They also used various means of trapping the bird, as in the use of pitfalls. The emu is a bird with a sense of curiosity, and the hunters made use of this fact by hiding in pits and making weird sounds, causing the birds to come in for a closer look. They also pounded pitori leaves with water, then baited the emu. The mixture stupefied the bird, thereby making it easier to spear.

The early Australians also ate the eggs of emus, and no doubt used the feathers.

They are known to have made knives from the long leg bones.

(See also BIRDS; OSTRICHES.)

eulachon (*Thaleichthys pacificus*) A member of the smelt family, the eulachon lives in the Pacific Northwest, from Northern California up to Alaska. The name came from an Indian word meaning "candlefish," which it is often called. The fish is very oily and the Indians fitted it (after it had been dried) with a wick made from the inner bark of a cedar tree. The wick soaked up the oil and, believe it or not, the fish could be burned like a candle.

The fish was also highly prized by all the coastal Indians in the Pacific Northwest, where it has been sold commercially. Although small and very oily, it is delicious when fried or smoked. Eulachon are also salted.

Although it swims upstream to spawn, the eulachon is essentially a saltwater fish and most of the ones sold in markets are taken in nets. In the sea, the fish feeds on tiny crustaceans and can't be taken by hook and line. During the spring spawning run, however, sport fishermen catch the tasty eulachon in large numbers in dip nets. This practice became so popular that regulations have been placed on dip-netting in American and Canadian waters, thereby limiting the numbers that can be taken.

The eulachon can be cooked in a number of ways, but it is at its best when grilled over a campfire beside the stream where it was caught. Because it is so oily, it doesn't dry out over hot coals and doesn't need basting.

(See also FISH.)

F

fish This popular term includes a number of species and countless subspecies. Many of these, such as eels and skates, don't really look like fish. Others look like fish but are really aquatic mammals, such as the porpoise and the manatee.

Most fresh- and saltwater fish are edible, but of course some are not as good as others. Some taste bad if not properly handled. Others are bony. Still others have soft flesh that spoils easily, although they might be delicious if they are taken out of the water and put into the frying pan. Some are good when cooked in one way but not another. Further, what is or isn't a good fish to eat depends partly on local custom and habits. A mullet, for example, might be highly prized along the coast of Florida, but 200 miles away in Louisiana it might be considered trash.

market fish Usually, these are caught in nets or traps, but some are caught more or less by hand. In some inland impoundments, for instance, many fishermen earn their living by catching catfish on trotlines. Large fish, such as swordfish and salmon, are caught by both commercial and sport fishermen.

The cods and herrings are very important commercial species, and have been for a long time, especially in North America and Europe. The herrings are often canned under the names kippers and sardines. The value of such fish to mankind would be hard to overestimate, especially in their salted form.

In the market as well as in restaurants, it is sometimes difficult to tell what's what. Shark meat has been marketed as gray fish, swordfish, and scallops. Skates and rays—both perfectly good table fare—have also been marketed as scallops. The tilapia has been marked as Nile perch. The list goes on and on, and this is apparently legal because it helps fishermen sell some fish that would otherwise not have market value. Whether or not a fish is marketable tends to change over the years, partly because of better methods of handling the catch and partly because the public changes its perception if what is or is not good to eat. Not long ago (when most fish were fried in American homes), most sharks were considered to be trash fish and were not eaten; but the recent emphasis on grilled fish has changed this somewhat, and now the sharks are being exploited perhaps too heavily. The same is true of the amberjack.

sport fish Some of the better fish for eating, such as the largemouth black bass, are not available in the marketplace. Others, such as salmon, are popular both as sport fish and as market fare. For eating purposes, some of the better fish are called panfish, a general term that includes a number of fresh- and saltwater species. These are usually small enough to fit into a regular sized frying pan. The term sunfish is rather misleading, since it contains the three-finger-sized pumpkinseeds and the largemouth bass, which grows to 20 pounds or larger.

Generally, the sportsman has a much wider selection of fish than is available at the market. In some cases, excellent fish that are caught by sportsmen are not eaten because of prejudice, lack of knowledge, or fear. Rays and skates, for example, are often caught but not used by saltwater anglers who are afraid of them because of the name stringrays or electric rays. While caution is certainly advised, most of these fish are not used frequently enough. What is a prize for one sportsman can be considered trash for another. In Europe, the carp is highly prized for both food and sport, whereas in the United States it is almost universally looked down on as both food and sport.

Probably the brown trout is the world's most important sport fish. A native of Europe, it has been established in Africa, South America, New Zealand, North America, and other places.

farmed fish Fish culture is very old, and was practiced centuries ago by the Chinese, Egyptians, and Romans. Today, fish culture is important in Asia, America, and Europe. Often fish culture is combined with another kind of farming, such as in rice fields that are flooded with water. Often, small fish are caught from the sea and put into ponds to mature. Eels are farmed in this manner, and in almost all cases the fish are not hatched in the rearing pond. Rather, they are raised or captured elsewhere and moved to rearing ponds.

In America, catfish and trout are currently the most important species for fish farming. Some connoisseurs, however, maintain that wild fish from free-flowing streams are usually better than pond-raised fish.

toxic fish Although most fish are safe to eat, a few such as puffers have toxic innards or roe that can be highly poisonous. Also, some predatory fish (such as largemouth bass) can become contiminated by mercury and other pollutants, and some predators in tropical waters (such as barracudas) can have the deadly poison, ciguatera.

(See also ALASKA BLACKFISH; ARMORED CATFISH; ARUANÁ; ATLANTIC CUTLASSFISH; BARRACUDAS; BARRAMUNDI; BASS; BONEFISH; BROWN TROUT; BURBOT; CABEZONES; CARP; CATFISH; CICHLIDS; CLIMBING PERCH; COELACANTHS; CONGER EELS; CRAYFISH; CUTTLEFISH; EELS; ELECTRIC EELS; ELEPHANT FISH; EULACHON; FLATFISH; FLYING FISH; GALAXIIDS; GARS; GOOSEFISH; GOURAMI; GRUNIONS; GRUNTS; GUDGEONS; INCONNU; JEWFISH; JOHN DORIES; LUNGFISH; MILKFISH; MORAYS; MULLETS; NILE PERCH; NOODLEFISH; OILFISH; PADDLEFISH; PIRANHAS; PIRARRA; PIRARUCÚS; PUFFERS; RAYS AND SKATES; RICE FISH; SAND EELS; SCORPIONFISH; SEA CUCUMBERS; SHARKS; SPINY DOGFISH; STURGEONS; TARPON; TOADFISH; TOTUAVAS; TRIPLETAILS; WHITEBAIT.)

flamingos These beautiful wading birds, pinkish in color, make up four extant species in family Phoenicopteridae. When feeding, the flamingo wades in shallow water, often on mud flats, and stirs up the bottom with its feet. Then it sort of dips up the small organic matter in its curved bill. (Actually, the top part of the bill ends up on the bottom and acts as a scooper and strainer.) The bird has a large tongue, which prevents it from swallowing large chunks of food.

At one time, the American flamingo could be seen in huge flocks along the Atlantic and Gulf coasts, but now most of the wild flamingos live in Africa and South America. The alkaline lakes of East Africa are favored haunts of both the lesser flamingo and the greater flamingo. In the past, the flamingo has been eaten, but its flesh is not too toothsome and its use as food was more or less a matter of necessity by poor local African peoples. In ancient Rome, however, where flamingo tongues were considered to be great delicacies, they were served at emperors' banquets.

(See also BIRDS.)

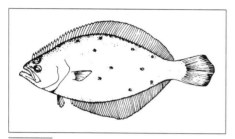

Flounder

flatfish These remarkable fish, ranging in size from small panfish to giant Atlantic halibut (*Hippoglossus hippoglossus*), and weighing up to 700 pounds, are all bottom feeders. Indeed, they hug the bottom sideways to such an extent that the fish swim sideways; in fact, both eyes are on one side of its head. When the flatfish are born, however, they swim upright and have an eye on either side of their head. As they develop, they start hugging the bottom to hide themselves from prey and, gradually, one eye migrates over the top of their head to the other side. Usually, the upper side of the fish takes on a mottled gray or brown coloration and the flat bottom side is white. Sometimes flatfish will completely bury themselves in sand, with only their eyes showing.

In all, there are about 500 species of left-eyed and right-eyed flatfish—including flounders, soles, halibuts, plaices, and flukes—in the world's seas and, sometimes, in fresh waters. All are good eating and some, such as the summer flounder (*Paralichthys dentatus*), are superb, having firm, white, mild-flavored flesh. Summer flounder and other species are also very popular with anglers. They are caught with live bait, cut bait, and artificial lures, and some are taken at night on shallow flats by foragers armed with spotlights and spears. Some of the flatfish are also important commercially.

(See also FISH.)

flies The term "fly" is a common name that is often used to describe many species of winged insects. There are over 60,000 species of "true flies," and countless others that are called flies. To say exactly which ones of these have been eaten by man would be impossible, but below a few are described that have been consumed on a regular basis.

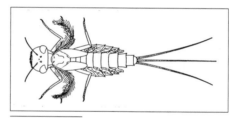

Mayfly nymph

Snipe flies, *Atherix ibis* and *Atherix variegata*, have a habit of laying their eggs on vegetation or limbs that hang over water. After laying the eggs, the female dies while sitting on the limb. More females come, sitting on top of each other, until a mass is formed. Being in clusters, they can be gathered in large numbers for food. According to *Insect Fact and Folklore*, the Modoc Indians of California gathered them by placing logs across streams, then shaking the bushes upstream. Of course, the dead flies fell into the water and floated downstream until they were stopped by the logs, where they could be dipped up in large numbers. As the book said:

As many as a hundred bushels a day could be secured in this way. The Indians used a basket to dip the flies from the water and to carry them to their ovens, where they were cooked. They were not taken out of the oven immediately, but were allowed to cool gradually. The Indians called this dish "Koo-chab-bie." When cold, it was about the consistency of headcheese and was firm enough to be cut into slices with a knife.

As all anglers know, mayflies also hatch from the water in large numbers. They are sometimes caught for food in Mexico and elsewhere. Even the mayfly nymphs, which live in the water, are sometimes caught by humans and eaten as food. Dragonflies are also eaten, but they don't swarm or gather in large numbers like the mayflies. Dragonflies do, however, have a habit of sitting on the end of a stick out over the water. Some peoples smear a sticky substance on the end of poles that entangles the dragonflies' feet, making them easy to catch. This trick, by the way, has also been used by primitive peoples to catch birds.

Even mosquitoes are eaten in tropical Africa, and were reportedly consumed by the Aztecs, who also ate water flies, larvae, and winged ants.

(See also INSECTS; MAGGOTS AND BOTS.)

flying fish About 100 species of flying fish (family Exocoetidae) live in the warm seas of the world. All make delicious table fare. Although they are rather small (from 6 to 18 inches long and quite slender), they are highly prized in some parts of the Caribbean and elsewhere. Usually, they are fish of the open seas and frequent areas that are rich in floating seaweed. They usually stay near the surface of the water, and have the ability to "fly"—no doubt developed as a means of escaping dolphins and other predators that feed near the surface.

The fish don't actually fly. They glide. All of the species have very long pectoral fins which, in flight, look and function exactly like fixed wings. When getting up speed to "take off," they first fold their wings tight against their body and get up speed with their tail and other fins. After breaking the surface of the water, they open their wings and glide without further means of propulsion. Up to a point, the larger the fish, the farther it can "fly." Some species have been known to glide for 1,000 feet.

Because they fall onto the decks of boats at night, some people believe that they are attracted to lights. Indeed, the fish have, reportedly, flown into the open portholes of ships! But it could well be that they are merely blinded by the light, not attracted to it. In any case, fishermen often use electric lights or torches at night while netting the fish.

Since it feeds on tiny fish and crustaceans found around floating debris or seaweed (especially sargassum), the flying fish is seldom caught on hook and line while bottom fishing or trolling the seas. A few have been caught by fly-rodders casting tiny flies, but it would be incorrect to suggest that this is, or will ever be, a popular sport. Most of the flying fish that are caught are taken commercially with the aid of nets in one way or another. They are usually fried, but they can be prepared in many ways.

In addition to feeding around floating seaweed and debris, flying fish also lay their eggs on floating matter. Exploiting these habits, fishermen off the coast of India are known to troll bundles of brush through the water. The flying fish are attracted to the brush, and are thus lured over large dip nets.

(See also FISH.)

fowl See BIRDS.

foxes All 21 species of foxes are edible. They live in just about every country in the world, from the hot deserts to the arctic regions. For the most part, foxes hold their own and are killed mostly for pelts, predator control, and rabies control. The fox is also hunted with hounds for sport, in which the chase is the essential element. The animal is not widely used for food, although it has been eaten in time of need, especially in the Arctic. The world's smallest species—the fennec fox, *Vulpes zerdu*, which weighs only about 3 pounds—is eaten in the Sahara. A beautiful animal, it has pale, sandy-colored fur that blends in with the desert setting, long ears and a long bushy tail.

There are reports of foxes being eaten here and there around the world, but the descriptions are usually vague as to species. Here's an account by Marco Polo from the kingdom of Fu-chau:

> Fox-like animals roam the whole area, damaging the sugar cane by gnawing at it; they also creep up on sleeping travellers and steal whatever they can find. The merchants have devised a method of trapping them. In the tops of large gourds they make an opening just large enough for the animal's head. To strengthen the opening they pierce holes around it and thread the holes with a piece of cord. They then put some fat inside the gourds and place them around their camp. The animals smell the bait and greedily force their heads into the gourds, but are then unable to withdraw them. As the gourds are very light the animals can lift them and walk off with them on

their heads. But they are unable to see where they are going so the merchants can then catch them easily. Their flesh is delicious to eat and their skins fetch a high price.

In any case, the ancient Chinese are known to have eaten both wolves and foxes, as well as domestic dogs.

frogs Some gourmets consider frogs' legs (meaning the hind legs of frogs) to be the best eating in the world. Many more ordinary people, however, know that the whole frog is edible—not just the hind legs. Indeed, the best part might well be the smaller front legs, whose meat, made up of short muscles, is not stringy. In any case, several of the world's 3,494 species of frogs and toads are eaten on a regular basis—to the tune of several million pounds per year. In Europe, the edible frog (*Rana esculenta*) takes its common name because it is widely eaten. In the United States, the American bullfrog (*Rana catesbeiana*) as well as other smaller frogs is commonly eaten.

Most of the frogs that are sold at market and eaten in restaurants are taken by market hunters in various parts of the world. Since frogs grow slowly and tend to be territorial, they are difficult to farm at a profit. (They also eat each other.) In the United States, frog hunting is a popular summer night's pastime, but it is now highly regulated by the various state fish and game commissions, some of whom require the practitioner to purchase a fishing license, whereas others require a hunting license. In Louisiana—a great state for frogging—the law doesn't allow the hunter to puncture the frog's skin, which rules out spears, gigs, and bullets; the frogs must be caught by hand or with a mechanical frog grabber.

Usually, frogging for meat or sport brings up an image of a swamp or marsh, such as the Everglades, but many of the world's great meat frogs are taken from more rugged terrain. In Africa, a giant frog (*Rana goliath*) lives along cold mountain streams and is quite wary and hard to hunt; such frogs grow up to a yard long (from nose to toe) and weigh up to 7 pounds. In South America, the natives of the Andes hunt the giant leptodactylid (which resembles a bullfrog) and roast it whole. In the West Indies, a large frog that grows on Dominica and Montserrat is an island delicacy called "mountain chicken" or crapaud. The Asian bullfrog (*Rana tigrina*) and other species are also widely eaten and are grown commercially in rice fields for export. Turkey also exports frogs.

Some frogs produce large and numerous tadpoles, and these were eaten by the Aztecs and no doubt by other peoples. Strangely, one small South American frog (which also lives on Trinidad) produces a giant tadpole that reaches 10 inches in length—about four times the length of the mature frog.

G

galaxiids When Captain James Cook set out on his third voyage to the Pacific regions (from 1776 to 1780), his mission was, among other things, to "observe the nature of the soil, and the produce thereof; the animals and fowls that inhabit or frequent it; the fishes that are to be found in the rivers or upon the coasts, and in what plenty. . . ." He might well have found plenty of small fish called galaxiids, so named because they are covered with so many gold-colored spots that they resemble a galaxy of stars. Most of these fish are small, but they do exist in large numbers in New Zealand (where Cook's naturalists obtained specimens) and around other large land masses in the southern seas, including Australia, South America, and so on.

The Maoris of New Zealand caught the small fish for food, and the early settlers from Europe followed their example. The Europeans called them whitebait. Today, the number of galaxiids has declined, but they are still caught commercially and are considered to be gourmet seafood.

Several species inhabit the southern seas, and most are migratory. Some of the seagoing species migrate up freshwater streams to spawn, like most of the salmon and trout of the Northern Hemisphere. In fact, the galaxiids are sometimes considered to be the salmanoids of the Southern Hemisphere. A number of species live here and there, and one of the most widespread, *Galaxias maculatus*, is found in Australia, Tasmania, New Zealand, Chile, Argentina, the Falkland Islands, and other places in between.

(See also FISH; WHITEBAIT.)

gars These primitive fish, often seen nosing about in shallow water, grow in North and Central America and in Cuba. All of the gars are slender and have a long snout with fine teeth. In fact, they are difficult to catch with a hook, but can be caught with a fishing line on which a strip of frayed cloth or felt (or velcro) has been attached. The gar's teeth become entangled in the cloth long enough for the angler to land it by one means or another. By far the largest of the seven species, the alligator gar (*Atractosteus spatula*) weighs up to 300 pounds. It prefers quiet waters in the lower Mississippi River system, but it also lives in a few other streams that flow into the Gulf of Mexico from Florida to Texas. Some anglers fish especially for the alligator gar with heavy tackle and wire leaders, and many others hunt the smaller gars with bow and arrow. This technique is called bowfishing, and the arrow is attached to the bow with a monofilament line that feeds off a spool. Bowfishing is encouraged in some states, not only for gar but also for carp and other so-called rough fish.

At one time, the alligator gar was smoked in Arkansas, but none of the gars are eaten except here and there on a regional basis. Reports from Texas indicate that fillet of gar is good when it is blackened, as in the famous blackened redfish dish from New Orleans, which became something of an American restaurant fad during the 1980s. In Louisiana, the gar is the main ingredient in garfish boulettes, which are sim-

---

Gar Balls Toogoodoo

Dress the fish and flake off the meat as described in the entry on Gars. If this flaking method is not used, it is best to grind the flesh in a meat grinder.

2 pounds (908 grams) garfish meat
1 pound (454 grams) boiled potatoes, mashed
2 large onions, finely chopped
1 cup (about 227 grams) parsley or celery tops
flour
mustard sauce
grease for frying
salt and pepper

Mix all of the gar meat, potatoes, onions, parsley (or other greens, including green onion tops, celery tops, or fresh chives), salt, and pepper. Form the mixture into balls about the size of a Ping-Pong ball. Roll in mustard sauce and then coat with flour. Deep-fry for a few minutes. Drain and eat.

—Adapted from *The South Carolina Wildlife Cookbook.*

---

ilar to codfish cakes. Garfish balls are also eaten in other areas.

In general, the gar has a soft flesh that is permeated with gristle. For the best care of the meat, shortly after the gar is caught it should be dressed—no easy task, considering its armored scales—and the meat should be put on ice. According to an account by James M. Bishop (with a recipe for longnose gar in *The South Carolina Wildlife Cookbook*), it's best to cut the fish with a heavy knife or hatchet along the top of the back, then peel the skin off both sides with the aid of a skinning knife. The flesh should then be removed with a fork or spoon. The idea is to scrape away the meat, leaving the gristle behind.

In any case, beware of the green roe of all gars. It is highly toxic to man,—although it makes excellent bait for bluegill. (See also FISH.)

gazelle These smaller cousins of the antelope exist over most of Africa, across southern Europe, and into Mongolia. Their habitat varies from desert to forest, depending on the species. All of the 18 species are edible, and modern methods of hunting from jeeps has endangered some species of gazelle.

Fortunately, one of the best species for the table, Thomson's gazelle (*Gazella thomsoni*)—a well-known grazer—is still quite abundant on the grassy flatlands of Tanzania and Kenya, partly, no doubt, because it can run at speeds of up to 40 miles per hour. According to Waverley Root's *Food*, "Thompson's gazelle, whose meat, prepared in the African manner by being marinated in berries and wine before being roasted with a basting of banana gin, is said to be of intoxicating excellence. . . ."

(See also ANTELOPE.)

gerbils These small rodents thrive in the deserts of Africa and the steppes of Asia. Marco Polo reported that the Tartars ate gerbils in the summer, when, he said, they were plentiful. He could have been referring to one (or to all) of the several species of gerbils or jirds. One species that lives in the area, the Mongolian gerbil (*Meriones unguiculatus*), tends to live in large social groups, which might make it seem to be more plentiful; this species is also the gerbil that is commonly sold as a pet. All together, 81 species of gerbils live either in Africa or in the Asian steppes.

Marco Polo also said that the peoples in a very cold land (which had white bears) ate gerbils, but most reference books don't place these rodents that far north. What he saw might have been a pika, which looks like a gerbil without a rodent's tail. The pika (a lagomorph, as is the rabbit) lives in the Asian steppes north of the Himalayas and on to Siberia, all the way to the Bering Strait. All together, there are 14 species of pikas, all of which weigh about half a pound each, and none of which lived in Polo's home grounds of Europe. The hamster also looks like the gerbil, and they live in pretty much the same areas as the pika. All three animals—the gerbil, the hamster, and the pika—are edible.

(See also HAMSTERS; RODENTS.)

giant spider crabs (*Macrocheira kaempferi*) As their name implies, all spider crabs have long legs. Often they are sluggish creatures and tend to disguise themselves from predators instead of making a speedy retreat; to this end, they camouflage themselves with sea plants and sponges. Spider crabs come in several sizes. The giant spider crab, or Japanese spider crab, is the largest of the many crustaceans, including the lobster. It inhabits the southeast coast of Japan in waters from about 100 to 160 feet deep. Since they have poor balance, these crabs avoid strong currents and much prefer to scuttle along a sandy or mud bottom under still waters.

Believe it or not, these crabs measure up to 26.5 feet between the tips of the legs. In the crab's normal defensive position, the claws are 10 feet apart. While they aren't especially dangerous or aggressive, such claws can, of course, inflict serious wounds. The crabs are eaten by anyone lucky enough to catch one, but they are not plentiful and are not of commercial importance. Called Takaashigani in Japan—meaning the tall-leg crab—the species was first reported in Europe by Engelbert Kaempfer, a doctor in the employ of the Dutch East India Company, who published a history of Japan in 1727.

The Tasmanian giant crab—which sometimes weighs as much as 30 pounds—is eaten in Australia, but the much smaller Queensland mud crab is more highly regarded as food.

(See also CRABS.)

giraffes These long-necked creatures live in the open woodlands of Africa south of the Sahara. Designed for feeding on trees, they are the tallest of all mammals. The larger specimens have an overall height of as much as 18 feet. The giraffes are heavy as well as tall, with the males weighing up to 2 tons. Color schemes vary with subspecies, but mostly they are made up of irregular brownish patches separated by cream-colored lines. The giraffe was known to the Romans, and Julius Caesar brought the first one to Europe in 46 B.C. The Romans called the animal "camelopardalis," which explains the last part of its scientific name: *Giraffa camelopardalis.* The first part comes from the Arabic *zarafa,* meaning "one who walks swiftly."

The giraffe provides very good meat for a lot of people, and may even be farmed on a limited basis. *The Encyclopedia of Mammals* reports that:

> Giraffe meat has for long been sought by subsistence hunters. Certain tribes, such as the Baggara of southern Kordofan in Sudan, the Missiriaas of Chad, and the Boran of southern Ethiopia, have traditionally hunted giraffe from horseback, the quarry receiving a special reverence and being of particular social significance to these tribes. But this mystique has not prevented local exterminations of giraffe by overhunting. The tourist trade in giraffe-hair bracelets has also encouraged poaching, only the tail being removed from the carcasses.
>
> Giraffes have important potential as a source of animal protein for human consumption. Because they feed upon a component of the vegetation that is unused by existing domestic animals, except possibly the camel, they do not compete with traditional livestock. Cattle ranchers and pastoralists have therefore been tolerant of giraffes, although they can damage stock fences. Giraffes rapidly become used to the presence of man and, as a free-range resource, can be easily cropped. At high densities, it has been estimated that giraffes could provide up to one-third of the meat requirements of a pastoral family, by exploiting the high browse which is currently neglected. This would reduce the dependence upon the sheep and goats that are responsible for much of the overgrazing and erosion problems in Africa today.

These long-legged animals also provide one of the greatest delicacies in all Africa: giraffe bone marrow. After the bones are baked in an oven, the marrow is picked out.

goats A surprising number of goats are raised for milk production, especially in cold and hot climates. (In more temperate regions, the cow is usually better suited for milk production.) The goat provides milk for small families who don't need all the milk that would be produced by a cow or for some reason don't want to feed a cow. A number of kinds of cheese, such as feta, are made from goats' milk. Further, goats are raised for leather and wool, especially Angora and cashmere.

Although goats, both wild and domestic, are eaten all over the world, they are not usually as commercially important as beef, lamb, or pork. They are eaten mostly by farmers and rural folk, and by people who raise them primarily for milk or wool. Goats are eaten in some mountain regions simply because they are more suited to the terrain than cattle, camels, or domestic sheep. In some parts of Italy and Spain, the meat is dried in the fresh mountain air.

In biblical times, the goat was reckoned in the wealth of the Israelites. The milk was highly regarded, as shown by Proverbs 27:27: "And thou shalt have goats' milk enough for thy food, for the food of thy household, and for the maintenance of thy maidens." The meat was also eaten. In addition, goats' hair was used for cloth and the skin was used to make bags for water and wine. The goat was also used for sacrifices. In fact, the modern word scapegoat came from the day of atonement (Yom Kippur), when the sins of man were symbolically put upon a goat's head. Then the goat was sent into the wilderness. Such ceremonial uses of the goat do not often occur in modern times, but some tribes in the Cameroons use goats for dowries.

In any case, goat meat is similar to mutton, and can be quite tasty if it is butchered and handled properly before cooking. Often, the young goats are eaten. Sexually active males have a bad reputation as table fare, and most people who raise them for food castrate them while young and then fatten them for the table. Some people even prefer milk-fed goats, called *cabrito* in the American southwest.

Wild goats and goatlike animals are difficult to classify, and sometimes it is difficult to determine what's a goat, a sheep, or a goat-antelope. There are, for example, 26 species of animals that are classified as goat-antelope—and many of them are called sheep. In general, however, all of the goats tend to live on steep terrains instead of on the plains. The goats are all members of family Bovidae, subfamily Caprinae, order Artiodactyla. These animals include the American bighorn sheep, the wild goat, the mountain goat (American), the musk ox, the chamois, and the chiru. All are edible, and some are highly prized as food.

All of the domestic goats that live around the world probably came from the wild goat, or bezoar (*Capra aegagrus*), native to the region roughly from the Greek Isles to Pakistan.

goosefish (*Lophius piscatorius*) Found in both the North and South Atlantic, this strange fish, which grows up to 6 feet long, has a very large head and a wide mouth full of teeth. Its eyes are on top of its head. The goosefish is a voracious eater, and is known to have consumed fish, lobsters, cormorants, and ducks. A member of the anglefish family, the goosefish even has a modified dorsal spine (a mobile filament) atop its head that may be used as a lure to attract prey to within striking distance of its jaws.

The goosefish is not highly regarded as food in America, except by a few local anglers. (Perhaps Americans think it is just too ugly to eat.) In Europe it is in great demand and is sometimes called monkfish at the market. Its delicate flesh is firm, sweet, and mild, and resembles spiny lobster in texture and taste. (Reportedly, goosefish tails are sometimes sold as scampi.) Also, the fish's large head has cheeks on either side that are considered to be delicacies. In France, the goosefish is often used in the popular bouillabaisses and similar fish soups because its flesh doesn't flake easily.

In addition to anglerfish and monkfish, the goosefish is also called bellyfish, frogfish, and sea devil.
(See also FISH.)

gourami (*Osphronemus goramy*) This freshwater fish is raised for food in Thailand and other parts of Asia, partly because it makes excellent eating and partly because it can live in warm and even stagnant water. (It can even survive for some time out of the water.) It grows to a length of about 2 feet and weighs 14 pounds or more. The gourami is the largest species of the family Anabantidae, which also includes several small aquarium fish, such as kissing gourami (*Helostoma temmincki*).

grasshoppers According to Leviticus (11:22), four species of buglike creatures

A Windfall Harvest

Locusts and other swarming, migratory grasshoppers, a plague to farmers and herdsmen in many parts of the world, are a boon to food gatherers living in arid regions where game is scarce. During the mid-1800s Maj. Howard Egan, a Morman pioneer, described an Indian "cricket" hunt held near Deep Creek in northern Nevada. His "crickets" are large, dark-colored katydids, long-horned and flightless (*Anabrus simplex*).

He encountered a group of Indians who, for several days, had been digging five or six trenches, about a food wide and a foot deep and thirty to forty feet long, joined together at the ends and facing uphill. The Indians covered the trenches with a layer of stiff, dry grass on which the insects were feeding. During the hottest time of the day, the Indians split up into two groups, with men, women, and children in each, and each person holding a bunch of grass in each hand. Each party walked to a place some distance behind either end of the trenches, and then spread out in a single line. Swinging their bunches back and forth, they gradually drove the "crickets" toward the line of trenches. As they drew nearer they slackened their pace, to give the insects time to crawl between the grass stalks into the trenches. When all had been driven in, the Indians set fire to the grass held in their hands, and scattered it, burning, over the grass on the trenches, creating a big blaze and much smoke, through which the "crickets" could not crawl out. The squaws removed the toasted insects from the trenches, which were then half full of them, and transferred them to larger baskets, which they loaded on their backs, along with their babies in bent willow cradles. Major Egan saw one woman carry off some four bushels of the insects, along with her baby, to the camp three or four miles away. Unable to carry off all the insects in one trip, they returned several times until the trenches were empty.

—Carleton S. Coon, *The Hunting Peoples*

could be eaten by the Hebrews: "Even these of them ye may eat; the locust after his kind, and the bald locust after his kind, and the beetle after his kind, and the grasshopper after his kind." Since the locust is really a grasshopper, it's difficult to tell for sure exactly what was intended. In any case, the katydids are also considered to be grasshoppers.

There are some 10,000 species of grasshoppers, and they live almost everywhere. Some 600 species can be found in the United States and Canada. Most have two large legs designed for hopping—but there are some strange members of the family. One species in South America (*Marellia remipes*), for example, lives on floating aquatic vegetation. It freely swims under the water, using hind legs that have been modified by nature for use as oars. In Yellowstone National Park, large numbers of frozen grasshoppers (*Melanoplus spretus*) have been found in glaciers; reportedly, Grasshopper Glacier high in the mountains of Yellowstone provides large numbers of preserved grasshoppers as food for birds and fish.

Often called locusts, grasshoppers have been an important source of food for man in some parts of the world. In Africa, the natives eat them in a variety

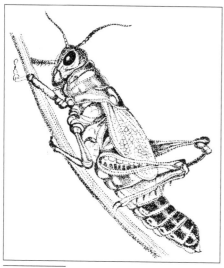

Grasshopper

of ways—raw, fried, roasted, boiled, and jellied. The grasshoppers are also salted and dried in the sun; later, they are eaten whole or mashed into a paste.

In terms of numbers, the migrating grasshoppers offer tremendous potential for local people to gather a food that is high in protein. Clouds of them often fly for long distances. Great hordes are found in North America, South America, Russia, Africa, West India, and other places. Even the desert has a grasshopper, *Locusta migratoria*, that swarms over North Africa, Syria, Iran, and West India.

Unfortunately, the swarming insects also do great damage to crops. Large swarms can eat every sprig of vegetation in their path, causing problems for people, livestock, and other animals. Swarms in the American west have been estimated at 124 billion grasshoppers. A swarm that once invaded Nebraska was 100 miles wide and 300 miles long; in some places, it was nearly a mile high. The agricultural damage caused by such swarms forced the Nebraska State Constitution to be rewritten; the revision became known as the Grasshopper Constitution.

In the Far East, grasshoppers have always been an accepted food, and they are eaten when found singly or in swarms. The American Indians also made good use of grasshoppers. The Digger Indians of California, for example, held great feasts when the grasshoppers swarmed. The insects were first soaked in salt water and then roasted in earthern ovens for a few minutes. They were eaten hot, or stored in a powdered form that could later be made into a soup.

Elaborate traps have been made—often involving ditches—for catching large numbers of migrating grasshoppers. They are also taken one at a time from meadows and other places when no migration is evident. Usually, they are easier to catch early in the morning when the dew is still heavy on the plants.

In some quarters, the dark juice exuded by grasshoppers is believed to cure warts. In both China and Japan the juice is sold in apothecary shops for medicinal purposes.

(See also CRICKETS; INSECTS.)

grubs and worms The exact classification of the various kinds of worms causes problems, but almost everybody understands the term in general. In any case, a number of wormlike creatures have been eaten around the world at various times.

Earthworms, including the famous nightcrawler (*Lumbricus terrestris*), are edible and have been considered excellent food in some lands. In France during the Middle Ages, epicures praised the taste of earthworms. In modern times, earthworms have been marketed as table fare even in America. The future of earthworms as standard American table fare remains to be seen, but the fact is that they contain 70% protein and they contribute in a large way to the diets of some people in South Africa, Japan, and New Zealand.

Earthworms live just about everywhere in tropical and temperate zones, and some are found above the ground in tropical rain forests. They also can be found in caves and under the bottom of lake beds. They do well in pastures, and some meadows in the Netherlands contain up to 7 million per acre. It has been calculated that often the worms under the pasture can weigh more collectively than the sheep above the ground. New Zealand is especially blessed with earthworms, and the Maori people had a custom of allowing only chiefs and dying men to eat certain kinds of worms. Some worms, it might be added, are large enough to provide a great deal of protein. Several species of giant worms grow here and there, such as the *Megascolides australis* of Australia which, when extended, may be 10 or 12 feet long. Earthworms are also called angleworms, manure worms, and so on.

Grubs have also been eaten extensively, especially by the aborigines of Australia, who eat witchetty grubs, and by some peoples of Africa. In fact, they are still eaten in Australia and Africa, as well as in China. According to Pliny, the ancient Romans even fatted grubs on flour before eating them. Again, however, the term grub is not at all scientific; it may include any sort of larval stage of various insects and may become confused with caterpillars. The silkworm, for example, is a great delicacy. Bots and maggots, both grublike, are also eaten. The popular term "grub" or "worm" would also include the agave worm (or maguey slug), which is still eaten in Mexico and was a great delicacy during the time of the Aztecs. Christopher Columbus reported that a white worm was eaten in the West Indies, and that it lived in rotting woods. Several grubs would fit that general description.

Caterpillars have sometimes been gathered for food in very large numbers. An account of once such harvest, as reported in *Insect Fact and Folklore*, follows:

> In the Cascade and Sierra Mountains of North America, Indian tribesmen seek the large caterpillars of the pandora moths which feed upon the needles of the yellow pines. The full-grown caterpillars measure from two to two and one-half inches in length and are as fat as an index finger. Normally, these caterpillars live on the pine trees, far out of reach, but in order to pass into the pupal stage they descend in great numbers to burrow into the soil. Just before this takes place the Indians build fires under the trees and stupefy the caterpillars by the smudge. This causes them to loosen their hold on the trees and fall to the ground where they are collected in baskets. They are then prepared as food by being dried over a bed of hot ashes or by being boiled in water.

Some edible worms live in the sea or in the sands or mud of coastal areas. These include clamworms, sandworms, and bloodworms as well as the arrowworms (Amphioxis). Arrowworms are not true worms but are nonetheless very good when eaten fresh or dried. Even the eggs of some sea worms are eaten (Palolo worm). Also, many fish have worms in their flesh; these are not only edible but are eaten unknowingly by very large numbers by people who do not normally diet on worms. For that matter, true vegetarians will have to look closely at their greens, tubers, and fruits; tons of worms and insects are inadvertently eaten each year. Also, it has been suggested that the European epicures who delighted in eating the trail of woodcock that had been properly hung were eating, for the most part, intestinal worms.

grunions (*Leuresthes tenuis*) Sometimes called California grunions, these small fish—about the size and shape of cigars—make excellent eating. The grunion is an unusual fish in that it lays its eggs on land, or at least in the sand. On selective high tides during spring or summer, usually at night following the full moon and the new moon, the female comes in with the surf, wiggles herself into the sand for 2 or 3 inches, and deposits up to 3,000 eggs. On the next series of highest tides, the eggs are uncovered and washed back into the sea, where the fry quickly hatch. Usually, this spawning activity lasts for only a few hours, so that thousands of grunion may be on a particular beach at the same time. Fishermen catch them with dip nets and by hand.

Also, the capelin (*Mallotus villotus*), a smelt of circumpolar distribution, also lays eggs in the sand at high tide, when they are caught by foragers with nets and seines.

grunts These small saltwater panfish, usually each the size of a man's hand, grow in shallow, warm waters and comprise 175 species—all of which are edible and quite tasty. The common name comes from the fact that when caught they grind their teeth together; this

sound is amplified by their air bladder, producing a gruntlike sound. One common species (*Orthopristis chrysoptera*) is called the pigfish or hogfish (because of the grunts) and lives from the Gulf of Mexico northward to Massachusetts. It is commonly caught from piers and small boats in the Carolinas by anglers using cutbait and flounder-fishing rigs. The flesh of the pigfish is mild, delicate, and grayish white. Small ones are usually fried, although they can be grilled and broiled.

The largest of the grunts, the white margate (*Haemulon album*), lives in the Florida Keys south to Brazil. Growing up to 2 feet in length, it is a fairly important commercial catch in the West Indies. It is good eating and is filleted more often than the smaller grunts.

gudgeons (*Gobio gobio*) These small fish live in most of the rivers of Europe. Although gudgeons prefer running water and are especially fond of riffles, they also live in lakes. Although their average length is only 5 to 8 inches, gudgeons are nevertheless an important species for anglers because they are plentiful and easy to catch.

The gudgeon is edible and is highly regarded in France, where it is usually fried. Its flesh has been described as sweet, delicate, tasty, and wholesome. At one time, the gudgeon was quite popular in England, and the catch-cook-and-eat gudgeon party was an annual event on the river Thames. These days, however, most serious anglers say that the gudgeon is best used as bait for larger fish.

The gudgeon resembles the stoneroller (*Campostoma anomalum*), often called hornyhead, which lives in streams in the eastern part of the United States.

It is delicious eating, particularly when fried, and is often caught by young anglers. Because it is so lively, it is also excellent live bait for larger fish. The stoneroller will take small wet flies, and it is sometimes stalked by fly fishermen in Tennessee.

The hornyhead chub (*Nocomis biguttatus*) is another small fish that resembles the gudgeon. It lives in streams in the eastern United States, grows up to 8 inches in length and is edible. The hornyhead chub is also used as bait for larger fish.

(See also FISH.)

guinea fowl Several species of this chicken-sized bird grow wild in Africa and are hunted for meat. The birds have been more or less domesticated, and for this purpose the *Numidia meleagris* species is probably the most popular, possibly because of its hooded head.

Guinea fowl are now raised in the West Indies, as well as in the southern parts of the United States. But it is a nervous bird that prefers to sleep in trees instead of henhouses. In short, it isn't suitable for large-scale production. It's a barnyard bird, and some people believe that they keep chicken hawks away. Other people keep them as "watchdogs," since they make loud noises when disturbed or frightened at night.

The birds may sometimes be found on the market, but in the United States this is more common in Louisiana than any other region. Some Creole or Cajun recipes can be found, along with a few from rural Florida. In general, the guinea fowl is cooked like barnyard chicken, which is a little darker and tougher than pen-raised chicken. The Romans called the bird the Carthaginian hen, and many rural people in the American South call it the guinea hen without reference to gender.

Guinea fowl have speckled feathers that are popular in the making of fishing lures. (See also BIRDS.)

guinea pigs (*Cavia porcellus*) A species of cavy that is widely used as laboratory animals and is highly popular as pets. The guinea pig no longer exists in its wild form, although a few feral animals no doubt live here and there. The guinea pig was domesticated as a source of food by the ancient Incas of Peru, Ecuador, and Colombia. The meat, which resembles pork, was considered to be a delicacy. The guinea pig was introduced to Europe as a food animal, not as a pet.

A rodent, the guinea pig feeds on grass and other vegetation. In captivity, it is often fed on dry pellets similar to those used for feeding domestic rabbits.

How the term guinea pig came about is open to speculation. During the slave trade after the discovery of America, European ships went to the West African country of Guinea before crossing the Atlantic, bound for the West Indies. Hence, the name guinea became associated with the slave trade. After delivering the slaves, the ships often went to the northeastern part of South America to pick up various goods to take back to England. Since the guinea pig had been domesticated long before the arrival of the Europeans, it is entirely possible that the sailors found some tame animals for sale along the east coast of South America, although the animals were originally from the Andes. Whether the name came from the African country by way of the slave traders is open to question. The German word *meerschweinchen* can be translated as "little pig from across the sea," and it is possible that the name came from Guiana in South America, so that its name was a corruption of "Guiana pig." In other words, Guiana might have been confused with Guinea.

Another explanation is that the animal, being small, could be purchased for a guinea. Because the rodent looked like a pig and tasted like a pig, that part of the name is easy to understand. Whatever its origin, the term guinea pig is now widely used to designate the subject or participant in an experiment, either knowingly or surreptitiously.

(See also CAVIES; RODENTS.)

gulls Often called seagulls, these birds are seldom seen very far out from shore. Nor are they restricted to the seaside or saltwater beaches. They are often seen far inland from the sea, especially now that more and more large hydroelectric dams are being built in the United States and other parts of the world. In any case, the gulls comprise 43 species. They live all around the world, except, strangely, for the islands between Australia and South America.

Most gulls are opportunistic feeders, eating the eggs and young of other birds, clams, insects, and garbage. The black-headed gull (*Laridae ridibundus*) feeds on insects on the fields of Europe. Other species are also attracted to freshly plowed dirt, where they can find grubs and worms. In America, Franklin's gull (*Laridae pipixcan*), sometimes called the prairie dove, feeds on flying insects. A smaller bird, the California gull (*Laridae californicus*), is credited with eating large numbers of grasshoppers (called Morman crickets) that threatened the first crops of the early Mormans. A statue of the California gull stands in Salt Lake City.

All species of gulls are edible, but the flesh has the reputation of being leathery. The eggs are another matter. Most gulls nest on the ground and are easy prey to foxes and other predators. That's why they nest on the offshore islands. But man can raid the islands in boats, and at one time gull eggs were taken by the untold thousands. Egging for market was common along the Atlantic coast until it was stopped by law. Egging still goes on in various parts of the world, but these days Atlantic gulls are in more danger from beach development than from egg poachers.

(See also BIRDS.)

H

hamsters These rodents live in Europe, China, the Middle East, and Russia, all the way to the Bering Strait. They usually eat vegetable matter, but will also feed on insects, lizards, and so on. They dig shallow burrows in the ground, which they fill with food for the winter. They are often considered to be pests in agricultural areas. Of the 24 species, the largest, at a pound and a half, is the common hamster (*Cricetus cricetus*) of central Europe and Russia.

Hamsters were eaten by primitive man, and are still used as food in China and elsewhere. They are often used as laboratory animals and are also popular as pets. The wild common hamster is hunted for its pelt, which, in cold regions, is sometimes used as lining for coats.

(See also RODENTS.)

hares See RABBITS AND HARES.

hedgehogs The common hedgehog, *Erinaceus europaeus*, lives in most of Eurasia. Eleven related species grow wild in Europe, Asia, and Africa. These tailless animals have sharp spines over all their bodies and, when frightened, simply roll up into a ball, which causes the spines to stick out in every direction. Thus, they tend to be difficult animals for predators to eat. It is said that the fox will sometimes roll the ball into the water and attack the hedgehog's vulnerable areas when it opens up to swim.

By day, the hedgehog seldom stirs from its den, usually a hollow tree, a hole in the rocks, or a burrow in the ground. At night, it feeds on worms, frogs, and so on, but it has also been known to eat fruit and other vegetation. Hedgehogs are usually considered to be an asset in rural England and other parts of Europe, where they are believed to help control mice and insects. Some people even put bowls of food out at night for "their" hedgehog to eat. Nonetheless, the meat of hedgehog is highly regarded as food and is reported to taste like suckling pig. In Britain, the gypsies are said to coat the whole hedgehog (spines and all) with clay and bake it in a pit. The spines are pulled off when the clay is cracked and removed, making the animal easy to eat.

hippopotamuses (*Hippopotamus amphibius*) These large mammals, which weigh up to 3,200 pounds, feed on terrestrial vegetation by night and stay in the water by day. Indeed, the hippopotamus can even walk along on the bottom of a stream or lake. The hippopotamus grows wild in east, central, and west Africa. A smaller species, the pygmy hippopotamus (*Choeropsis liberiensis*) lives mostly in Liberia and the Ivory Coast.

The fatty meat of the hippopotamus is very highly regarded by some Africans.

honey See BEES AND HONEY.

horses (*Equus caballus*) In all probability, the horse was first domesticated as a source of food, not as a beast of burden. In any case, horseflesh was commonly eaten by early peoples, and hippophagy, as the practice is called, still goes on in some areas, particularly in times of famine and war. Although the flesh was forbidden by Mosaic law, Joseph bought up horses to be used as food during a great famine. The French—especially Parisians—have always eaten horseflesh, although it has been outlawed from time to time. The butchers in Paris have even

---

Scythian Boiled Horse

The Scythian country is very ill-provided with wood, so they contrive to boil meat in this way. They flay the victim and strip the flesh off the bones and throw it into cauldrons, if they have any (the cauldrons of the country are much like the Lesbian cauldrons, except that they are much bigger); and they boil them by burning the bones. When they have no cauldron, they put the flesh in the beast's stomach mixed with water and boil it over the bones. These make excellent fuel, and the flesh, when it is taken off the bones, fits easily in the stomach; thus an ox cooks itself, and so do beasts of other kinds. When the meat is cooked, the sacrificer takes a first offering from the meat and offal and casts it on the ground in front of him. They sacrifice horses mostly, but other animals as well.

—Herodotus of Halicarnassus

---

held horseflesh banquets and launched advertising campaigns to promote the food.

The Teutons used horseflesh in their daily fare until they were converted to Christianity. The horse is also eaten in Asia and was a very important meat to the Mongols and other peoples of the steppes.

Mares' milk is drunk in some places, and the Mongolians use it to make *kumiss*, a fermented drink. At one time, Mongols made extensive use of blood as food which, of course, could be drawn without killing the animal.

Truly wild horses are now extinct, although feral horses, such as the Mustangs in the American West, still exist in good numbers. Several species of African zebra live in the wild and are edible, but these are separate species from the horse. Donkeys and mules are also eaten, and seem to be more highly regarded than horses as table fare.

(See also BLOOD; DONKEYS; MILK.)

horseshoe crabs Also called king crabs and swordtail crabs, these ancient creatures, which have been on the earth for 200 million years, grow to a length of 2 feet and live along the Atlantic coasts of North America (as well as in the Gulf of Mexico) from Maine to the Yucatan, and along the coast of eastern Asia. Their bodies are shaped like horseshoes and, unlike most crabs, they have ratlike tails. They can turn on their backs and use their legs to paddle along on top of the water. Horseshoe crabs are eaten from time to time, but in America most fishermen chop them up and use them for chum or to bait eel pots. They have also been used as fertilizer along the Atlantic coast. They lay eggs in depressions in the sand at high tide; these eggs are eaten in Asia.

(See also CRABS.)

hydraxes These animals are often confused with rabbits because they are similar in size, and hop about in the same manner. In fact, they are sometimes called rock rabbits. They are even prepared like rabbits. Hydraxes inhabit Africa south of the Sahara and in the extreme southwestern fringes of Asia. Some 40 million years ago, hydraxes were important grazing animals in parts of Egypt, and some of these extinct species were much larger than today's remaining species.

The modern hydraxes comprise 11 species and 3 groups: tree hydraxes, genus *Dendrohyrax*; rock hydraxes or dassies, genus *Procavia*; and bush hydraxes, genus *Heterohyrax*. Their popular names pretty much describe the habitat preferred by the various species, and all weigh between 5 and 12 pounds. All of the hydraxes are eaten for food, and the eastern tree hydrax is hunted for its fur in the forests that surround Mount Kilimanjaro. Curiously, the hydraxes always urinate in the same place, causing white mineral deposits to form. These crystals were used as medicine by some natives of South Africa.

According to *The Encyclopedia of Mammals*, in the Phoenician and Hebrew languages the name for the hydrax is *shaphan*, meaning "the hidden one." When the Phoenicians explored the western Mediterranean, they saw a large number of rabbits in what is now Spain. They thought they were hydraxes, and named the land "Ishapan" for Island of the Hydrax. This name was later changed to Hispania and, finally, shortened to Spain.

I

ibis These beautiful birds have long down-curved bills and long legs, designed for feeding in shallow water and marshes. One species or another can be found, in suitable wetland habitat, throughout the tropics and temperate regions of the world—except in the islands of the South Pacific.

The sacred ibis (*Threskiornis aethiopica*) was very important in the religion of ancient Egypt. Records of the sacred ibis go back 5,000 years. The birds were depicted in art, were a part of the hieroglyphics and were even mummified and buried with the pharoahs. The god Troth—whose duty was to record the history of every human being—is identified with the sacred ibis and is depicted with the head of the bird. The sacred ibis is now rare in Egypt, but it does live south of the Sahara and in Madagascar.

Because of its beauty, the ibis has been killed in large numbers by commercial feather hunters. Its meat is said to be oily, but it is nonetheless relished by some people who live near large populations of the bird. (See also BIRDS.)

iguanas Most of the 700 species of Iguanidae are native to the tropical regions of Central and South America. Most of the lizards are rather large, and some are quite unusual. One species is called the only marine lizard; although it seldom takes to the open sea, it does feed on seaweed at low tide. A species in Cuba can actually run across the water, then dive to the bottom to escape danger. Although many iguanas use water as a means of escaping danger, some are creatures of arid lands. One such iguana, the so-called horned toad of the American Southwest, is a lizard that, when frightened or angered, can squirt blood out of its eye for a distance of several feet.

Although all iguana are edible, only the common iguana (*Iguana iguana*) of Central and South America has achieved fame as a gourmet food. In fact, it is daily fare for the Indians, who grill the meat over open embers. The meat—white, tender, and of mild flavor—can also be cooked in many other ways, including fried, baked, etc. The iguana tail is considered to be the best part, but the rest of the animal is also quite tasty.

An arboreal creature, the iguana prefers to live in the jungles and forests, where it feeds on leaves and berries, as well as on an occasional bird or small mammal. Surprisingly, the iguana attains a length of up to 6 feet (including the tail). It is fond of lazing over the water on low-hanging tree limbs, and it won't hesitate to drop into the water when disturbed. From a boat, a casual observer might mistake it for a water snake or even a small alligator.

The iguana is such a good source of food that some countries, such as Nicaragua, have taken steps to increase its numbers and regulate its harvest.

inconnu (*Stenodus leucichthys*) These little known members of the Salmonids family live in Alaska, northwestern Canada, and Siberia. An isolated tribe also inhabits the Caspian Sea. Normally, the inconnu live in salt water and run up river to spawn, but some have become landlocked in lakes. Also called sheefish and connie, the term "inconnu" is French, meaning unknown. Like many other Salmonids, the inconnu is highly regarded as a sport fish. Although it isn't too interested in flies, it will often take spoons cast with spinning tackle. Once hooked, it is a great jumper, and in fact it is sometimes called the Eskimo tarpon. The fish can reach 55 pounds.

Opinions about the inconnu's merits as table fare are mixed. The flesh is oily and is best when smoked. In addition to those fish caught by individual fishermen for private use or for sport, large numbers of inconnu are taken in nets and sold commercially. The Eskimo eat the fish on a regular basis.

(See also FISH.)

insectivores The zoological order Insectivora includes 345 species of small animals, usually no larger than a rabbit, that live mostly in North America, Africa, Europe, and Asia. They don't live in most of South America, Australia, or Greenland. The larger insectivores are eaten (or have been in the past) wherever they are found—especially in Madagascar and the West Indies. The larger species include the tenrecs and the hedgehogs, discussed under separate entries.

Another family of the larger insectivores—the Solenodons, which look like American opossums with anteater snouts—live in the West Indies and have in the past been highly regarded as food. They were quite plentiful when the white man arrived. Unfortunately, both species (*Solenodon cubanus* of Cuba and *Solenodon paradoxurus* of Hispaniola) are now endangered, largely because they are eaten by the dogs, cats, and mongooses that have been introduced by man.

The various moles and desmans have been trapped extensively, not for food, but for fur and because they were considered to be pests. There may be evidence that shrews were eaten by ancient man, but these days most shrews are too small to be of much value as food. (The pygmy white-toothed shrew is in fact the world's smallest mammal, weighing only 0.07 ounces.) It is true that the elephant-shrews are eaten in Africa, but these animals aren't insectivores.

(See also HEDGEHOGS; TENRECS.)

insects Over 2 million species of insects live on Earth, and some of these are very important to man in that they play essential roles in nature, such as pollination of plants. Products from insects—honey, silk, wax, dyestuffs—are perhaps more important to man than the use of the insects as food.

Nevertheless, insects themselves are eaten and have played important roles in the daily diets of some primitive peoples, especially in Africa, South America, and Australia. Insects are widely eaten in Asia, both as gourmet fare and as more ordinary daily food. In some cases, large swarms of insects such as grasshoppers provide bonus food but are not part of the daily fare.

In Europe and North America, the insect is more likely to be a curiosity for gourmets than a serious supplement to the diet. Generally, insects are more commonly eaten in Mexico than in the United States or Canada. The "gusanos" marketed in Mexican delicatessens can be agave worms, caterpillars, earthworms, or beetle grubs.

(See also ANTS; BEES AND HONEY; BEETLES; CICADAS; COCKROACHES; CRICKETS; FLIES; GRASSHOPPERS; GRUBS AND WORMS; LICE; MAGGOTS AND BOTS; PALOLO WORMS; SILKWORMS; TERMITES; WATER BOATMEN.)

J

jabirus (*Jabiru mycteria*) These American members of the stork family range from Mexico to Argentina. They stand as high as 5 feet, and have a large bill. Their heads and necks have no feathers. In spite of declining numbers, the jabiru is hunted for food in parts of Central and South America. Other species of stork are also eaten in Cameroon and, according to Waverley Root in *Food*, storks were also eaten by Henry VIII during the 16th century.

jellyfish Believe it or not, an edible jellyfish living in the salt waters of eastern Asia (*Phopilema esculenta*) is highly prized in parts of China and Japan. Reportedly, it has been sold under the name of sea blubber in Hong Kong restaurants. The ancient Romans also ate jellyfish, although no details are available on the exact species that was consumed. Dried jellyfish can sometimes be purchased in Chinese markets.

jewfish (*Epinephelus itajara*) These large members of the grouper family, sometimes called Junefish or spotted grouper, live in the tropical regions of the western Atlantic, as far north as Florida—and in the West Indies. A similar species also grows in the Pacific from Costa Rica to Peru. Although the fish grows up to 700 pounds, it likes to hang out in relatively shallow water, usually around rocky shores, reefs, and bridge pilings. Indeed, the jewfish is often caught—or at least hooked—by anglers fishing from bridges. Many large fish are better when they are young, but the jewfish is good eating regardless of size. The meat is fine grained, white, and mild in taste. The flesh is best when fresh or frozen, but it has been salted and marketed as salt cod in the West Indies.

During its lifetime the jewfish undergoes a sex change. It starts life off as a female and ends up as a male.

Some of the other large sea bass and groupers are also called jewfish in some areas. These include the edible Warsaw grouper, which weighs up to 500 pounds, although the larger specimens are best used in chowder and should be skinned. (See also FISH.)

John Dories Nobody knows how this big-mouthed fish got its name. It makes very good eating and is available commercially in Europe, where it has been highly esteemed since Roman times. An American species (*Zenopsis ocellata*) can be found from Cape Hatteras to Nova Scotia, but it is not commercially important simply because it is not plentiful. Another species lives in the eastern Atlantic from England to South Africa, and it is available in European markets. Since the John Dory is mostly head, however, only about one-third of its weight is edible meat—a fact that tends to make it expensive. The fish has delicate white flesh, similar in taste and texture to the turbot.

The John Dory has a curious spot centered on either side of its body, and no one knows what its purpose is. This spot is sometimes called Saint Peter's mark. One explanation has it that Peter once caught a John Dory, grasped it between his thumb and forefinger, and then released it back into the sea. In any case, the fish is called *peterfisch* in German and *saint-pierre* in French. (See also FISH.)

K

kangaroos and wallabies These Australian animals have pouches for holding their young and long hind legs for hopping. Their front legs are short, almost like arms. When fighting, the males sometimes box each other. The kangaroos have large muscular tails, which they use for balance and which man uses to make an Australian dish called kangaroo tail soup. The red kangaroo (*Marcopus rufus*) is the largest of the group, weighing almost 200 pounds and sporting a tail 42 inches long.

Similar pouched animals include the wallabies and the smaller rat kangaroos (not to be confused with the kangaroo rats) of Australia and neighboring islands. During his 1770 voyage to the Pacific, Captain James Cook ran aground a reef and stayed in eastern Australia (now Queensland) for several days. While there, he collected three "specimens," the description of which was of great interest in Europe. Part of Cook's report about the specimens read: "Land animals are scarce, as so far as we know confined to a very few species; all that we saw I have before mentioned, the sort that is in the greatest plenty is the Kangooroo, or Kanguru, so call'd by the Natives; we saw a good many of them about Endeavor River, but kill'd only three which we found very good eating."

Of course, the aborigines also ate the kangaroo—as well as the wallaby and rat kangaroo—whenever they could catch or kill them. In Queensland the natives cooked kangaroo by filling the body cavity with hot stones, wrapping the body with bark, burying it in a hole over hot ashes, covering it with sand, and building a fire on top. In addition to using its meat, the aborigines also hunted the kangaroo for its hide. In fact, some of the women used kangaroo skins to make dresses—and carried their babies in the pouch.

The early European settlers also made good use of the kangaroo for both meat and hide. Before long, however, cattle and sheep provided more than enough meat for local consumption. Kangaroo is still eaten in the outback, where the legs are roasted or boiled, but these days the meat is used mostly for dog food. In any case, the number of kangaroos has been greatly reduced, partly because of loss of habitat to farming and to cattle or sheep ranching.

(See also MARSUPIALS.)

Kangaroo Tail Soup

Skin the kangaroo tail and cut it into joints. Heat a little oil in a Dutch oven. Chop 2 or 3 medium to large onions and brown them slightly in the hot oil. Add 5 pints (2.37 liters) of beef stock and 4 tablespoons (60 milliliters) of barley. Dice or chop 4 carrots, 2 turnip roots and 4 stalks of celery. Put the vegetables into the pot. Add 3 bay leaves and a little chopped parsley, if available. Bring to a boil, reduce heat, cover, and simmer for at least an hour, or until the kangaroo tail is tender. Stir in ¼ (59 milliliters) of good red wine and remove the pot from the heat. Let it sit for a few minutes before serving.

---

krill (*Euphausia superba*) These cold-water shrimplike crustaceans are small, but they are plentiful, especially in the Antarctic. They grow up to 3 inches long and feed on planktonic organisms. Krill also reproduce rapidly and form a dense reddish-brown layer in the sea. Like a blanket, this layer sometimes covers hundreds of square miles; it is usually about 40 feet deep, but it can go down to 3,000 feet.

This bonanza of food is exploited mostly by the baleen whale, which has been reduced in recent years by fishermen. Now the krill are being caught and used as food for man and livestock. Just how important this source of food will become to mankind remains to be seen. But the food is there for the taking—between 500 and 750 million tons—in the Antarctic.

L

land crabs Several species of crabs live on land, often at a considerable distance from water. These usually eat vegetable matter, and some even climb trees for their food. Most of these species are hermit crabs of the family Gecarcinidae. The true land crabs breathe air through a gill system, but they return to the sea to lay their eggs. Some of the land crabs live in the shells of land snails, and others live in holes in rocks or even trees.

Land crabs are especially plentiful in the West Indies and on the West Coast of Africa. In the Caribbean, they are strung together and sold at market. Before being brought to market, however, they are kept in pens for several days and fed a diet of coconut meat, corn, or scrap breadstuff. The meat of the land crab is different from that of the sea crabs—being a little more chewy—and this is part of the reason for fattening them. In any case, they are highly prized as table fare on some of the Caribbean islands. Land crab pilau is served in Tobago, and *crabes farcies* (stuffed land crab) is a popular appetizer in Martinique and Guadeloupe.

Apparently the crabs are quite plentiful on some islands, or have been so in the past, and they scuttle about quite rapidly and with audible clanking. According to Waverley Root in *Food*, in 1585 Sir Francis Drake and some of his men prepared to ambush Spanish troops in Santo Domingo. Drake mistook the clanking of a horde of land crabs for a large number of armored troops—and retreated back to his ships without doing battle!

lapwings (*Vanellus vanellus*) A plover, the lapwing ranges across northern Europe to Siberia and southward in summer to North Africa, India, and Japan. It also visits North America, but not in large numbers.

In the past, the flesh of lapwing has been highly regarded as food. In Europe, its eggs—called plover eggs—were in great demand and were marketed by the millions. As the demand was greater than the supply, the lapwing eggs brought high prices in French markets. Many of the old French recipes are for hard-boiled lapwing eggs and, surprisingly, they were not often served hot. (Even when hard-boiled, the albumen of these eggs is a milky color and remains soft.) In any case, the sale of lapwing eggs is now illegal in some countries and hunting is not widely practiced these days.

Although considered to be a wading species, the lapwings are mostly inland and upland birds. They usually nest in a shallow depression on open ground. Sometimes the depression is lined with straw or pebbles. The hen usually lays four eggs.

lemurs Ages ago, lemurs probably lived in many parts of Europe, North America, and Africa, but most of them seem to have retreated over the years to the island of Madagascar. In general, lemurs live in rain forests much like monkeys and are noted for their ability to leap through trees.

Several species of large lemurs have become extinct on Madagascar, partly because (it is believed) of the arrival of hunters on the island about 2,000 years ago. One of these early lemurs had a skull of over a foot long. The largest extant lemur, the indri (*Indira indira*), weighs up to 22 pounds.

The indri lemurs live almost entirely in trees. Being long-legged creatures, they can leap up to 33 feet. Of course, they depend on forests and have become threatened as the rain forest of Madagascar is cut for fuel, timber, and agricultural expansion. The indri is also hunted and trapped for meat by the local people.

lice There are several species of bloodsucking lice, and all are highly selective in that they thrive on a particular mammal. The gorilla louse (*Phthirus gorillae*), for example, doesn't care to live on the chimpanzee. Man is the choice for *Pediculus humanus*, a head louse, and a subspecies, the body louse. Another species, *Phthirus pubus*, prefers the pubic area instead of the head of humans.

The louse is a very old human pest, and specimens have even been found in ancient mummies in Egypt. It is the louse that carries the dread typhus fever as well as relapsing fever and trench fever. Typhus, however, isn't caused by the bite of lice. Rather, it comes from an organism known as *Rickettsia*—which is present in lice feces—and usually infects humans through cuts or scratched skin. Typhus has been particularly problematic during times of war. When the French besieged Naples in the 16th century, for example, a tragic 21,000 out of 25,000 men died of typhus.

The louse has figured in folklore and strange customs around the world. In England, the hog louse was used in a medicine called vinum millepedum—hog lice wine. In Newfoundland, anyone who was infected with lice was considered to be a healthy person. This belief isn't too farfetched because the lice prefer to live within a degree or two of human body temperature. Lice will quit a person with high fever as well as a corpse when it begins to cool. Farther north, the Eskimo of Greenland, who had to while away the long arctic nights, made a sort of Igloo game of catching each other's lice. Anyone who was fortunate enough to find a louse let out a squeal of delight and ate it on the spot, alive. When available in large numbers, lice were sometimes saved for a special treat.

(See also INSECTS.)

limpkins (*Aramus guarauna*) A large rail-like bird of the marshlands, the limpkin is so tasty and easy to bag that the early settlers in South Georgia and Florida hunted it almost to extinction. Now that the bird is protected by law, its numbers have increased and it is fairly common in the area between the Okefenokee Swamp of Georgia and the Florida Everglades. The walking bird feeds mostly on small aquatic creatures at the water's edge, as well as on worms and small frogs. In the Everglades, its diet is made up largely of freshwater snails. It has a sharp 4-inch beak and measures up to 28 inches in length. The limpkin's feathers are grayish brown, speckled with white, and the bird blends in well with reeds and marsh grass.

The name "limpkin" came from Florida, probably from the bird's limping gait. It is also called the courlan or crying bird because of its wailing call, usually heard at night or late afternoon. Its toes are long, designed for walking over muck and vegetation. The limpkin can and does swim nicely, but it does not fly often and has to make a short run in order to take off. Once it is in the air, it resembles the crane, with its neck extended and its feet dangling. Limpkins are not migratory birds. They do sometimes feed in woodlands, and they usually roost in trees at night.

Four subspecies of the limpkin also live in Central and South America, from Mexico on down to Argentina, as well as in the West Indies. They are prepared in dozens of ways, but are usually fried.

lizards At least 7,351 species of lizards live in the temperate regions of the world. Some live as far south as Tierra del Fuego, and one species survives as far north as the Arctic Circle in Norway. Most lizards live on land (or in trees) but a few live in water, or freely take to the water for pleasure, to escape predators, or in search of food. A few species of lizards look like snakes and have no legs.

Most of the lizards can be eaten for survival purposes. Perhaps the iguana is the best known edible lizard, at least in the Americas. It is standard fare in Central America, South America, and parts of the West Indies, although it has become more difficult to find in the wild. In India, monitor lizards are eaten on a regular basis, and are also used for leather. Even the giant Komodo dragon, or ora monitor (*Varanus komodoensis*)—a threatened species—has been hunted for food. These monitors grow to a length of 10 feet and weigh up to 360 pounds. Other monitors almost as large live in Australia, where they are also eaten. The aborigines coated lizards with mud before roasting them in hot coals from an open fire. All monitors lay long soft-shelled eggs, which are also edible. They often lay the eggs in holes along river-

banks. The Nile monitor deposits its eggs in termite nests.

In addition to being hunted for meat and robbed of their eggs, large lizards are sometimes killed for their skins, which are used in fancy leather goods.

(See also IGUANAS.)

llamas (*Lama glama*) These smaller members of the camel family were domesticated in the Andes mountains long ago and were used extensively by the Incas. In addition to their use as pack animals, they also provided food, wool, and leather. Even their dung was used for fuel. Today, no llamas live in the wild, except possibly for a few feral animals here and there. Further, the numbers of domesticated llamas have also declined. The animals are not often used for wool these days because the sheep serves this purpose better. Also, because of better roads, trucks, jeeps, and railroads, their use as a pack animal has declined greatly. But they are still used as pack animals in the Andes and elsewhere, partly because, like their camel cousins, they don't require much water. A good llama of 200 pounds can carry a load of 130 pounds for up to 20 miles per day. If they are overloaded, however—or forced to travel too far without rest—they will lie down, kick, and hiss. (The camel will also fuss and bite if it thinks it is being overworked.) Recently, llamas have been introduced into various parts of the world, including the United States, as pets and as pack animals for recreational campers and backpackers.

Although the llama makes good eating and has meat similar to beef, its use for food in the old days was deemphasized simply because it was so valuable as a pack animal.

The llama has three long-necked cousins that still live wild in South America. The largest of these, the guanaco (*Lama guanicoe*) weighs up to 265 pounds. It is not a mountain animal like the llama. Nor can it be domesticated. It lives from the foothills of the Andes in Peru down into the grasslands and scrublands of Argentina and on into Tierra del Fuego. The Ona depended on the wild guanaco for food and hide. The animals were used for their fur, for roofing, and for such leather goods as moccasins. Also, the bones were used for tools and weapons. At one time, guanacos were present in large numbers, and Charles Darwin, who partook of the flesh, is said to have seen 500 guanacos together on Santa Cruz. Usually, however, the animals prefer to stay in small herds of 15 or 20. In recent years, the animal has been hunted commercially for its pelt. This species is now protected in both Chile and Peru, where it is not as plentiful as in Argentina.

Like the larger guanaco, the reddish orange vicuña (*Lama vicugna*) also refuses to be domesticated. It lives high in the Andes of central Peru, Bolivia, Chile, and Argentina. Its meat is quite good, but it has been hunted even more for its wool, also called vicuña, which is of a fine texture and makes valuable woolens. Because of declining numbers (from the millions when the Spanish arrived in South America to less than 15,000 in 1960), the animals are now protected by law.

The alpaca (*Lama pacos*), like the llama, now exists only in the domesticated state. It has been used as food for centuries, but its main use has been as a source of wool. The alpaca has long rather shabby hair, and wool made from it is light and strong with good insulating properties. It is used in the manufacture of sleeping bags, coats, and so on. Long ago, the Incas reserved the use of alpaca wool for royalty.

(See also CAMELS.)

lobsters The true lobsters live only in the salt waters of the northern part of the Atlantic Ocean. Two species are usually recognized: the American lobster, *Homarus americanus*, which lives from Labrador to North Carolina, and the European lobster, *Homarus gammarus*, which lives from Norway to the Mediterranean. Both species are so similar that many people believe them to be variations of the same one. In any case, the lobster is highly regarded as table fare on both sides of the Atlantic. Of course, the meat from its large claws is eaten as well as that from its tail. Some epicures, especially in Europe, maintain that the best parts are the "coral"—or premature eggs of the female—and the green liver.

Lobsters are caught in traps or pots by commercial fishermen, and these days the catch is highly regulated so that only lobsters of a certain size may be taken. The color of live lobsters varies from greenish blue to reddish brown. When cooked, the lobsters turn pink. The meat is sold canned or frozen, and the lobsters are also marketed fresh—alive, in tanks—and are even shipped alive in damp seaweed. Although large lobsters are sometimes caught, most of them average about 2 pounds.

The spiny lobster (*Panulirus argus* and other species) is a different animal. The tail part looks like a lobster, but the head doesn't have the large claws. Usually, only the tail part is eaten, and the meat is highly regarded. On average, they are a little smaller than the true lobsters, although they can attain weights of up to 17 pounds. Several species of the spiny lobsters grow in the warmer salt waters from North Carolina south to Brazil. The "rock lobsters" and similar cousins are harvested commercially along the coast of Africa, Australia, California, and in the Mediterranean. All of the spiny lobsters and rock lobsters, since they live in warmer waters, are more popular among sport fishermen, who often hunt them with the aid of skin- and scuba-diving equipment.

Several smaller kinds of decapod crustaceans are edible. The Florida lobsterette *Nephropsis aculeata*, for example, is smaller than the spiny lobster. Several species live around the world, including the North Atlantic, and all are excellent eating. They are sometimes marketed under such names as Danish lobsters or scampi. Norway or Dublin lobsters are still another species of small lobster, and these are called scampi in Europe. All of the lobsterettes have firm, sweet flesh that is highly prized in restaurants. The delicious Murray lobster of Australia is really a large freshwater crayfish, which grows up to 10 inches in length.

Lobsters are usually cooked by boiling or steaming.

(See also BULLDOZER; CRAYFISH.)

loons (family Gaviidae, genus *Gavia*) Four species of loons live around the artic regions from Greenland to Siberia, sometimes flying as far south as Florida. They are almost totally aquatic birds, and are noted for their diving prowess and their ability to swim long distances underwater. They have solid bones, so that their specific gravity is almost the same as that of water. Measuring from 2 to 3 feet in length, loons are quite awkward on land—from which they can't get airborne. Even on water, they have to "run" for up to a quarter of a mile before they can take off. Once airbone, however, they are strong fliers and can go up to 60 miles an hour. They are wide-ranging birds. The Arctic loon (*Gavia arctica*), for example, was found, at one time or another, in Scotland, the Bering Sea, Hudson Bay, California, China, Japan, and the Black Sea. Loons are beautiful birds, and have a wailing call that is sometimes mistaken for the howl of a wolf. Although they have the reputation of eating lots of fish, they are protected in many countries.

Loon meat is said to be fishy and tough, but some peoples have developed a taste for them. American Indians and the Eskimo used the skins to make clothing. The breast and neck feathers are used for decorative purposes. The bird is sometimes taken by hunters, but this is mostly a regional sport. The natives of Harkers Island, North Carolina, were once called "loon eaters" because they had a reputation for hunting and eating the birds, which they often stewed with rutabagas.

(See also BIRDS.)

lungfish This primitive group of freshwater fish can live on air and even bury themselves into the ground during dry weather. Some of the several species (such as *Protopterus dolloi* of the Zaire basin) burrow into the soft ground head first, then turn head up and become encapsuled as the mud hardens and their bodies become cocooned in a bodily secretion. They can live on their own fat and body tissue for three years if necessary.

The lungfish currently lives only in the Amazon basin, parts of Central and West Africa, and Queensland in Australia. Their fossils, however, have been found in rocks as far north as Greenland and as far south as Antarctia.

The fish are eaten in Africa, where they can sometimes be dug up during dry periods.

(See also FISH.)

M

maggots and bots Around the fringes of the Arctic, a species of botfly, *Hypoderma lineata*, deposits its eggs on the backs of caribou. When the eggs hatch, the larvae burrow through the skin so that they can feed on the tissue. As the larvae grow, boillike swellings called bots occur just under the skin. When the larvae reach full size, they again eat their way through the skin and fall to the ground, where they hatch.

Some of the Indians in the Hudson Bay area of Canada use tame caribou for meat and for beasts of burden, just as the Lapps do. The Dogrib tribe learned to squeeze the swellings on the caribou, forcing the bot to pop out. The Indians ate them alive. When an infested caribou was slaughtered for the table, the Dogrib simply left the bots alone and cooked them along with the meat.

Maggots are also eaten, dead or alive, in other parts of the world, especially when very large animals, such as elephants, are infested during butchering.

(See also GRUBS AND WORMS; INSECTS.)

mammals From mice to elephants, all land mammals are edible, and a good many species are raised especially to feed the human race. Others, large and small, have been hunted and trapped for food. Primitive man hunted even very large mammals, such as the extinct mastodons and the huge whales of the sea.

In many cases, animal fat, hides, milk, blood, and bones were also very important to mankind. Oil from whales, seals, and manatees, for example, was used for fuel, light, and heat before electricity was invented. Even domestic animals have been valuable for oil (such as pork lard), and the fat-tailed sheep of the Middle East were bred especially for producing fat.

(See also AARDVARKS; AARDWOLVES; AFRICAN POUCHED RATS; ANTEATERS; ANTELOPE; 3ARMADILLOS; BABIRUSAS; BADGERS; BATS; BEARS; BEAVERS; BISON; BLOOD; BONES AND BONE MARROW; BUFFALO; CAMELS; CANE RATS; CAPYBARAS; CATS; CATTLE; CAVIES; CHAMOIS; CHIMPANZEES; CHINCHILLAS AND VISCACHAS; CHIRUS; CIVETS; DEER; DEER MICE; DOGS; DOLPHINS; DONKEYS; DORMICE; DUIKERS; ECHIDNAS; ELANDS; ELEPHANTS; ELEPHANT-SHREWS; FOXES; GAZELLE; GERBILS; GIRAFFES; GOATS; GUINEA PIGS; HAMSTERS; HEDGEHOGS; HIPPOPOTAMUSES; HORSES; HYDRAXES; LEMURS; LLAMAS; MAN; MANATEES AND DUGONGS; MILK; MONGOOSES; MONKEYS; MUSK DEER; MUSK OXEN; MUSKRATS; NARWHALES; NUTRIAS; OKAPIS; ORANGUTANS; OTTERS; PANGOLINS; PECCARIES; PIGS; PLATYPUSES; PORCUPINES; PRAIRIE DOGS; PRONGHORNS; RABBITS AND HARES; RACCOON; RATS; REINDEER; RHINOCEROSES; RODENTS; SEALS; SKUNKS; SLOTHS; SPRINGHARES; SQUIRRELS; TAPIRS; WAPITIS; WARTHOGS; WHALES; YAKS.)

man (*Homo sapiens.*) It's true. Throughout history—and even into modern times—man has fed on man either of necessity or by design—and even by organized hunting. More often than not, modern cannibalism is a measure of last resort, as in survival on a raft or a case of distress in remote regions, such as the Arctic. In some instances, cannibalism has been practiced in the form of rites as, for example, when eating part of a felled warrior in the belief that his vitality will be passed on to the victor of the battle. This belief was common among the so-called headhunters.

With a Touch of Salt

Recipes for *Homo sapiens* are hard to find, but the Aztecs had a dish called *tlacatlaolli*, or maize and man stew. Maize, of course, is American corn, but merely mixing these two ingredients will sound rather bland to accomplished gourmands. Fortunately, there is evidence of more ingredients being used in the Aztec way of cooking up Spanish explorers. When advancing on the great city of Tenochtitlán, Cortés noted that the Aztecs—in thankless return for the Spanish coming with the commands of God and the king—had already prepared boiling pots of tomatoes and peppers (both of which were New World ingredients) as well as a touch of salt. Unfortunately, no exact measures for the ingredients were recorded.

In another type of cannibalism, a piece of flesh was cut or bitten out of a living person, which would not necessarily mean death. Blood was also drawn from living people as well as the dead. In Australia, a few tribes considered cannibalism of the deceased to be a demonstration of respect. But in some cases the main purpose of cannibalism was meat, and humans have been an important source of food. In fact, the Batak people of Sumatra actually sold human flesh in the market until the Dutch took over and stopped the practice. In the Congo living men, women, and children were bought and sold for meat. In the Solomon Islands, captured people were fattened for food. Further, some cannibals had firm opinions on the quality of the meat, and often women were preferred to men. In good cheer, a chief in Tahiti said that the white man, when well roasted, tasted like a ripe banana, whereas other islanders held that white sailors were too salty and tough, much inferior to a Polynesian.

Marco Polo also reported cases of cannibalism in China. Regarding one area, he said, "The people here eat revolting things, even human flesh if the man has not died of an illness. If he has been killed by, for instance, a sword, they eat all of him and claim that the meat is excellent." Polo also said that the people on some Japanese islands regarded human flesh as the greatest possible delicacy.

The term "cannibalism," by the way, came from the Spanish word for Carib— a people of the West Indies—who were notorious cannibals.

manatees and dugongs These large aquatic mammals live in tropical and temperate waters. All of them feed on grass or vegetation that grows in the water. Growing up to 1,900 pounds, manatees make excellent eating, and the smaller dugongs are also highly regarded as food. Although both are called sea cows, they taste more like pork. In fact, the dugong is sometimes called the sea pig, but this is because of its habit of rooting up the bottom in search of food. In Portugal the manatee is sometimes called *peixe-boi*, meaning fish cow. In addition to the meat, manatees and dugongs have in the past been hunted for their hides and for oil, which in some regions was used for cooking and for lamplight.

The dugong (*Dugong dugon*) lives in the coastal shallows of the Western Pacific and Indian oceans and often roots up the rhizomes of certain sea grasses, as indicated above. The adults have tusks,

Mother and infant manatees

and these have been of value to peoples who hunted the dugong for food, hide, and oil. These animals have been associated with accounts of mermaids and other myths around the world.

The manatees comprise three species—the Amazonian manatee (*Trichechus inunguls*), the West Indian manatee (*Trichechus manatus*), and the West African manatee (*Trichechus senegalensis*)—all of which live in coastal areas as well as in suitable fresh waters. The West Indian manatee also lives in Florida, where it is fully protected by law. Part of the danger in Florida waters and elsewhere is from high-speed boats, which slash the rather sluggish manatees with their propellers.

All of the manatees are considered to be vulnerable, due to heavy hunting and loss of habitat. One interesting new use of the manatee, however, is to keep grass out of irrigation canals and heated waterways associated with power plant discharge.

An extinct cold-water species called Steller's sea cow (*Hydrodamalis gigas*) grew quite large—up to 13,000 pounds. It lived in the shallow waters around a few subarctic islands in the North Pacific and fed on kelp. The animal was discovered in 1741 by a shipwrecked Russian exploration party, and word of its good eating and easy capture caused seal hunters to camp on the islands to get meat. As a result, Steller's sea cow was hunted to extinction by 1768. At one time, it roamed the coastal waters from Northern California to Japan.

mantis shrimps (family Squillidae) About 350 species of these viscious marine predators (not true shrimp) inhabit the shallows of the world's warmer seas. The mantis shrimp is rather elongated and has huge front pincers or claws, giving it the appearance of a mantis in the water. It burrows into the sand, or hides in crevices, while waiting for a small fish or some such creature to come within reach of its claws.

Mantis shrimp aren't normally seen in fish markets because they are difficult to capture in large numbers. The shrimps grow up to 12 inches long, however, and are sometimes caught by foragers after food. They can be caught with a wire snare baited with a small fish, mollusk, or crustacean, but the snare must be placed in the burrow. In any case, mantis shrimps are excellent eating, and the long tails are boiled or fried like regular shrimp.

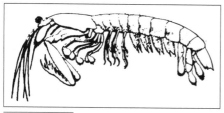

Mantis shrimp

marsupials This order of mammals comprises 18 families and 266 species. A few of them (the opossums) live in the Americas, and the rest can be found in Australia, New Zealand, and islands in the Indo-Pacific area. At one time, a few members of the marsupial order lived in Europe, but these disappeared long ago. The animals are very old, dating back at least 75 million years in North America and 23 million years in Australia. In any case, all of these mammals give birth early, and in most species the young are reared in pouches.

Primitive and modern men have made extensive use of marsupials. The obvious examples are the kangaroos of Australia and the opossums of the Americas. Lesser-known animals include the brushtail possums (very common in Aus-

tralia) and the cuscuses, the larger species of which have been important to local peoples for their hides and for meat. Other edible marsupials of the Indo-Pacific region include the bandicoots, koalas, ringtail possums, and wombats.

(See also KANGAROOS; OPOSSUMS; WOMBATS.)

milk Universal baby food for all suckling mammals—as well as of baby whales, kangaroos, and other teatless species—milk is not widely exploited except by man. Although man has made extensive use of milk from other mammals since recorded history, its use is by no means universal. Many American Indians, Africans, and Asians of all sorts—according to Waverley Root—cannot digest milk properly. Further, some adults consider the idea of drinking milk to be repulsive.

But many people do drink milk, and this trend will likely continue now that modern dairy practices and processes have been developed. For many years, dairy farming and processing has been an important industry in Europe and America. Today, the term milk almost implicates the cow, which has been bred to produce more and more milk on a daily basis. But cow milk may not be the best kind for human consumption. Goats' milk is both richer and easier to digest than cows' milk, and is often prescribed by doctors. Asses' milk is closer to that of humans, and therefore might be better for our babies to drink. Sheep's milk is sometimes drunk in the Sahara and other places where sheep are an important part of the culture. It was also the favorite milk of the Romans, who descended from shepherds and sometimes mixed it with wine. Today, however, both sheep's and goats' milk are probably more widely used in cheese than as a beverage.

Much of the milk consumed in India comes from the buffalo, an animal that is widely milked in other parts of Asia as well as in parts of Africa. Camels' milk is, of course, frequently used by the Arabs and possibly by other nomadic peoples of the desert. In Tibet, the yak fills the need. In Lapland, the reindeer gives a rich milk, which contains three times as much protein as cows' milk and seven times as much fat.

But the earliest and perhaps the most extensive use of milk was on the Eurasian steppes, where the horse was all-important not only for transportation and meat but also for milk. Even today mare's milk is used to some extent—and is even brewed into a spirited beverage called kumiss in Mongolia and among other nomadic peoples of Asia. (Kumiss is also made from camel's milk.)

The use of cheese and butter is also very old, and they were probably used as food more extensively by nomadic peoples in Asia than in other parts of the world. Cheese is mentioned in the Bible, and butter in the Hindu Vedas. Neither the Greeks nor the Romans used butter, except on their hair and on skin injuries. The use of concentrated milk goes back to the time of Genghis Khan and was used as food during military missions. A patent for concentrating milk was issued in the United States to Gail Borden in 1856, and such milk was in fact used as army rations during the Civil War. A British patent for dried milk was issued during 1855, but the product was not used widely until the turn of the century.

In addition to cheese, other milk products include whey, kefir, sour cream, and yogurt.

milkfish (*Chanos chanos*) Although the milkfish is not well known in Europe or the Americas, it is quite important in the Indo-Pacific regions, where it is also known as bangos and bandeng. The fish has no teeth and feeds mostly on algae, which reduces its value as a sport fish to almost nil. It does, however, grow up to a length of almost 5 feet and a weight of almost 50 pounds.

The mature milkfish as such is not very important in the catch of commercial fishermen. They do, however, catch and market untold millions of milkfish fry from coastal waters. These fry are purchased for raising in brackish water ponds, which are usually shallow (about 2 feet deep) and are often constructed by building dikes in tidal flats around the mouths of rivers. This permits the aquaculturists to maintain the best balance of fresh and salt water for maximum production of algae and milkfish. It is a highly developed kind of farming, and is widely used in some areas. Java, for example, produces millions of pounds of milkfish each year in ponds, some of which have been reclaimed from mangrove swamps.

Adult milkfish are, of course, sometimes caught from the ocean. They range from East Africa, across the Indian Ocean, on into the Pacific and eastward to Hawaii and central California. (See also FISH.)

mongooses (family Viverridae) One or another of 31 species of these animals live from the Atlantic coast of Africa across southern Asia and into the Philippines, and they have been introduced elsewhere. Although best known for their ability to kill snakes, they also feed on insects, small vertebrates, and even fruit. One Asian species (*Herpestes urva*) even feeds on crabs. A species from India was loosed on the Hawaiian Islands and in the West Indies during the late 1800s in an attempt to control rats on sugarcane plantations. Before long they were considered to be pests, and were quick to attack chickens.

Some of the species tend to be abundant in their range, and all have no doubt supplied food for local peoples at one time or another. Marco Polo reported that the Tartars regularly ate an animal called Pharaoh's rat, said to be abundant in the steppes. It was probably a kind of mongoose. At any rate, the animals have been eaten during modern times in Chad and South Africa.

Mongooses are easily tamed, and they make good pets. The ancient Egyptians considered them to be sacred, and their images have been found in tombs and temples dating back to 2800 B.C. Modern man's image of the animal is shaped, at least in part, by Rikki-Tikki-Tavi, the fictional mongoose that did battle with a cobra in Rudyard Kipling's *The Jungle Book*.

monkeys In parts of Africa and Asia, monkey isn't merely a survival food. It is the favorite meat of many of the natives, and even in recent years it has been sold in markets and restaurants. The meat is also highly regarded in parts of South America and, reportedly, monkey has recently been served in a restaurant in Grenada. Gorilla has also been served in Gabon. In addition to monkeys, apes, baboons, gorillas, and lemurs are also eaten. All of the 181 species of extant primates are edible, although, of course, some species may be more toothsome than others.

Some primitive peoples developed special techniques for trapping and hunting monkeys, as shown in the following from Carleton S. Coon's *The Hunting People* in which the Birhor people of India, who live on the remote Chota Nagpur plateau, conclude a net hunt:

> Once the monkeys have been killed one of the hunters kindles a fire by twirling a bamboo shaft on a hearth board. The hunters then singe the bodies, cut them up, and remove the hearts, brains, and other organs to be cooked and eaten on the spot, but not at once. First the chief roasts a little of the meat, and offers it to the spirits. After the men have eaten their snacks, the chief divides the rest of the meat according to fixed protocol, and the hunters and beaters carry their shares home.

Coon also points out that the Birhor people are quite at odds with some of their Hindu neighbors, who believe the monkey to be sacred. Also, the use of the monkey as food has caused cultural conflicts in recent times. According to Waverley Root's book *Food*, the monkey "caused strained relations between Zaire and Egypt when hotel cooks in Alexandria refused to cook for the Zaire soccer team the monkeys it had foresightedly brought along with it to provide proper provender for their athletes in a country which they had correctly expected might be deficient in it."

Although monkeys have been very important food even in modern times, some people, knowingly or unknowingly, partake of species that are endangered and should be protected. The lion tamarin, *Leontopithecus rosalia*, is a good example. These beautiful monkeys are almost extinct in the wild. Native to parts of Brazil, they are restricted to low-altitude forests, habitat that is steadily being destroyed for timber and agricultural expansion. The monkeys feed on fruits, frogs, lizards, snails, and other small animals. Some of the lion tamarins are trapped and exported for zoos—and some are eaten by local Brazilians.

All of these monkeys have manes like lions, and some are of a beautiful red color. One subspecies, the golden lion tamarin (*Leontopithecus rosalia rosalia*), is especially handsome. Another species, the golden-rumped lion tamarin (*Leontopithecus rosalia chrysopygus*), may be the rarest monkey in South America. (See also ORANGUTANS.)

morays Some 120 species of these ill-tempered, eellike fish live in the tropical, subtropical, and temperate regions of the seas. They prefer to stay in shallow water around reefs and usually live in a cave or crevice. Poor swimmers, they are not often spotted in the open seas. Often morays will ambush an octopus or other suitable creature from their holes, sometimes with their heads sticking out a foot or two. They are highly protective of their holes, and are aggressive toward anything that approaches. Morays have been know to attack skin divers who come too close to their hiding place. Most species grow up to 10 feet long and have sharp teeth. Although they like to live in a hole, they do move about, and some even enter fresh water.

The oily flesh is eaten in many parts of the modern world, but it is not highly acclaimed. The ancient Romans, on the other hand, were quite fond of the moray. In fact, the fish got its name from a wealthy Roman epicure by the name of Licinius Muraena. The species common to the Mediterranean is *Muraena helena*. Some of the Romans actually raised morays in ponds connected to the Mediterranean by canals. A few Romans are said to have made pets of the creatures and even decorated them with jewels. The reports of slaves being fed alive to the morays, however, are no doubt incorrect. Although the moray will bite if threatened, it must swallow its prey whole—in one piece—in order to eat it. The moray's teeth are not designed for chewing or for biting out chunks of flesh, as piranhas do. On some species, the sharp teeth are even hinged so that a rather large creature can be swallowed. But the moray simply isn't large enough to swallow a man.

Morays are sometimes speared or hooked by fishermen, but they are difficult to land or boat. They twist and turn and even tie knots, in themselves and in the lines, as they attempt to tear out the hook or break the line. Morays have even been choked and killed by the knots they made in fishing lines and wire leaders. American eels also tie knots in themselves in an effort to escape from hooks, but morays are probably more adept at this tactic because they use it often when feeding on octopuses. (The moray first grabs the octopus by the head; the octopus immediately wraps its tenacles around the moray. The moray then throws a loop with its tail and works itself through it, thereby ridding its body from the octopus's tenacles. Once free, the moray quickly swallows another inch or so of the octopus, and the process is repeated until the end.) Even when boated or beached, the moray doesn't give up. It can bite like a dog and strike like a snake.

(See also EELS.)

mullets Over 100 species of mullets swim and jump in the world's tropical and temperate waters, but identifying them quickly leads to confusion. In one noted reference work, for example, the mullet is classified as two kinds: red mullet and gray mullet. The gray mullet make up the family Mugilidae, of which the more common and widespread species (growing in the Mediterranean as well as on both sides of the Atlantic and in the Gulf of Mexico) is the striped mullet, *Mugil cephalus*, called the black mullet in Florida.

The confusion does not stop with the color. There are widely differing opinions on the quality of the fish as food. In Florida, the mullet has a very good reputation indeed and many tons are brought to market each year. In parts of Louisiana, on the other hand, they are difficult to give away, much less sell at market.

Most reference works, however, agree that mullet has a fine, delicate flesh that makes good eating. The trick to succulent mullet, coastal peoples say, is to eat it as soon as possible after taking it from the water. The fish is often fried, but is best when smoked or grilled. In Florida, freshly caught mullets are butterflied and grilled over charcoal, or mangrove wood, without being skinned or scaled. At one time, salted mullets were an important staple in some areas and were sold from wooden boxes open to the air. Salt mullets are still widely available, although today they are usually wrapped in plastic film and refrigerated. In the southern part of the United States, salt mullets are sometimes eaten for breakfast with grits. In any case, mullet is usually oily and is extremely high in mineral content, especially iodine.

The mullet's reddish yellow roe is very good and is available (from Florida fish) in the fall or early winter. The roe can be cooked in a number of ways. Some coastal people in the fishing villages along North Carolina's Outer Banks, as well as other rural places along the sea, salt the roe and dry it. Dried salt roe is also eaten in some areas around the Mediterranean. *Batarekh*, as it is called in Egypt, was eaten by the pharaohs.

The real gourmet treat from the mullet, however, is the soft roe or milt, which has a mild flavor and a delicate marrowlike texture. Going even further, the culinary adventurer may want to look into the mullet's innards. The fish have a very long intestine track (used to digest its diet of vegetable matter) along with an organ that functions like a gizzard. In fact, it can be dressed and cooked just like a gizzard from a chicken or other fowl.

Millions of pounds of mullet are marketed each year by commercial fishermen, who take them in long nets. Most of these fish weigh a pound or two, but five-pounders are not uncommon. Since mullets feed mostly on tiny vegetable matter, they are not often taken by sport fishermen. For a long while it was believed that they could not be caught by hook and line. In recent years, however, knowledgeable anglers have learned to take the mullet on light tackle with tiny salmon egg hooks baited with an okra seed or perhaps with a tiny ball of bread. Many fishermen catch them for sport and food with the aid of circular cast nets. Since mullets sometimes swim in large schools, they are also taken by snatching or fowl hooking with weighed treble hooks attached to a line.

Mullet is usually taken close to shore and is common in bays and inlets. Most species frequent brackish water, and some swim for considerable distances up freshwater streams. One minor species lives in fresh water. In addition to their use as food, small mullets are frequently used as bait for larger sport fish, such as tarpon. (See also FISH.)

musk deer (family Moschidae) Three species of musk deer live in Asia, generally from northern India to Siberia. Smaller than most of the true deer, the musk deer reaches a maximum weight of about 37 pounds. Its meat is eaten, and the young ones are just as good as any other venison—the name musk notwithstanding.

The animals are, however, hunted and trapped more for their musk than for food. This substance, produced by a gland in the male, is used in making expensive perfume as well as in medicines. In India it is used in aphrodisiacs and stimulants. There are several commercial grades of the musk, and the better stuff is said to be worth its weight in gold. The traffic in the musk is considerable. In a single year, Japan imports 5 tons of it. Since each musk pod weighs about an ounce when dried for market, many male animals are required in order to obtain musk in quantity. Unfortunately, musk deer tend to use well-established trails through the woods, which makes them susceptible to snares and traps. For this reason, many of the does are killed along with the musk-bearing bucks. In recent years, however, steps have been made to remove the musk without actually killing the animals. (See also DEER.)

musk oxen (*Ovibos moschatus*) Shaggy-haired ruminants of the far north country, these animals are really a type of goat antelope instead of a true ox. A million years ago, the musk ox lived all around the Arctic Circle and as far south as France. Gradually, its range contracted—possibly from hunting—until it was limited to the arctic areas of North America, mostly in Greenland and east of the Mackenzie River delta. It is an animal of the arctic barrens, feeding on sparse grass, willows, lichens, and other vegetation on the tundra near glaciers.

The musk ox is a large animal, standing 6 feet at the shoulders and weighing up to 1,400 pounds. Its meat and milk are of excellent quality. It can be cooked like beef and other good red meat. The Eskimo hunted them not only for meat, but also for their hide and hair, which were used for clothing. Musk oxen are shabby creatures, and their dark brown hair hangs almost to the ground.

The animals like to travel in groups of 20 to 30. When attacked by arctic wolves and other predators, they form a circle with the larger animals on the outside, head out, presenting a rim of horns. But such a defense is no match for modern firearms, and the musk oxen have been hunted almost to extinction. They have made something of a comeback, and have even been transplanted to some other arctic areas, such as Nunivak Island in the Bering Sea.

muskrat (*Ondatra zibethicus*) Neither the musk nor the rat part of these animals' common name is very appetizing to modern peoples. Nonetheless, the meat of this semiaquatic rodent is quite edible. In fact, the animals can be very tasty, especially the young ones. The meat has been sold commercially in Maryland and Louisiana in recent years, sometimes under the name marsh hare.

Baked Freshwater Mussels

Recipes for freshwater mussels are difficult to find. Here's one from Dana Crosland in South Carolina, who says, "These are found in our rivers, creeks, and branches, usually underwater in varying sizes from tiny to big. I bring them home after duck hunting, choosing those that are about the size of our saltwater clams."

freshwater mussels
salt and pepper
Worcestershire sauce
fresh lemon juice

Wash the mussels well in fresh water and keep them very cold. When ready to cook, preheat the oven to 400°F (204°C). Mix the salt, pepper, lemon juice, and Worcestershire sauce. Arrange the mussels in a biscuit pan, add a little water, cover tightly with aluminum foil, and put them into the oven for a few minutes until the shells open. (If required, use more than one pan.) Take the pan out of the oven, and discard the aluminum foil and water. Switch the oven heat to broil. Remove the top of each mussel shell and return the rest to the pan (on the half shell). Pour a little of the sauce mixture over each mussel and broil the batch in the oven for 2 minutes. Serve on the half shell.

—The South Carolina Wildlife Cookbook

In the past, the muskrat has been one of the most important fur animals in North America, at least in terms of numbers and dollar value. At present, the pelt of the muskrat is used mostly as trim and as coat lining, but it has also been used for hats, gloves, and coats. In an effort to increase the popularity of the fur, muskrat has been marketed as river sable, Hudson Bay seal, and water mink. In addition to its pelt, the animal also produces a musky oil that is of value in the perfume industry.

The muskrat is native to North America, and can be found from the tundra in the north of Canada to Mexico. It has also been stocked in Europe. The muskrat lives in underground lodges near the water, or in mounds made of reeds and grass in shallow water. Inland, the rodents live along streams or the banks of lakes, often in beaver ponds. In some tidewater areas, they thrive in shallow marshes or grassy areas. Their lodges or mounds usually have an underwater opening, making them difficult to spot from land. Muskrats feed almost entirely on vegetation such as reeds and roots in or near the water. They grow to about a foot in length and have a long, scaly, ratlike tail. (See also RODENTS.)

mussels Various species of saltwater mussels, such as the *igai* (*Mytilus crassitesta*) of Japan's Inland Sea, are eaten around the world. The edible mussel or blue mussel (*Mytilus edulis*) of both sides of the North Atlantic has been an important food for the peoples of England, France, and other parts of Europe, but it is not highly consumed in America. Similarly, though the primitive peoples of Europe and the British Isles ate *Mytilus edulis* in large numbers, the American Indians did not, in spite of the fact that these bivalved mollusks thrived in the waters of New England and parts of Canada. Of course, the early American settlers from Europe paid close attention to what the Indians ate, and were therefore reluctant to eat *Mytilus edulis*, despite its similarity to European food. (The same mystery also arises with the soft-shelled steamer clam, *Mya arenaria*, which has always been highly regarded in North America, but not in Europe.)

In any case, mussels do have great potential as a cheap source of food for mankind. Although they are susceptible to man's pollution in the wild and readily transmit diseases from contaminated waters, mussels can be farmed and have in fact been raised in France as far back as 1235. According to *The Encyclopaedia Britannica*, it has been estimated that an acre of prime mussel-producing area can produce up to 10,000 pounds of flesh a year. By comparison, the best cattle lands will produce only about 150 pounds of beef.

To add to the confusion about what is edible and what is not, freshwater mussels are eaten by some peoples, but not by others. Although the coastal Indians of the northeastern United States did not eat saltwater mussels, the Osage of the Ozarks—as well as some other tribes—made good use of the freshwater mussels. Some authorities say that freshwater mussels are really clams, although they are called "mussels." Other authorities say that freshwater clams are really mussels, although they are called clams. By whatever name, many bivalved mollusks live in fresh water all around the world, especially in the Mississippi River and its tributaries in North America. At one time, Illinois alone had more than 70 species of mussels. In recent years, however, environmental changes caused by projects such as the channelization of

streams and dams have caused about 20 of the species to disappear from the state. At present, 29 species are on the endangered list in Illinois.

In the past, freshwater mussels have been gathered commercially in the United States, and Illinois has always been a leader in the harvest. During the 1800s, mussels were used to manufacture tons of pearly buttons. The button industry continued to grow here and abroad and, in 1932, some 54.2 million pounds of mussels were gathered nationwide by professional fishermen.

The advent of plastics pretty much killed the pearly button trade, although mussel shells continued to be used on a limited basis as "genuine pearl" buttons, as well as for items such as fishing lure spinner blades. The mussel fishing industry sprang up again when it was discovered that bits of mussel shells were ideal for implanting in oysters to encourage the growth of cultured pearls. For this purpose, Illinois alone exports more than 1,000 tons per year—mostly to Japan. The meat from these mussels is usually sold for fish bait or hog food.

Although primitive peoples used large quantities of freshwater mussels as food, modern man doesn't relish them like he does oysters and other bivalved mollusks. Over the years, various proposals and studies have set forth for making better use of freshwater mussels as table fare. So far, nothing much has come of such efforts. Nobody seems to know exactly why, but they may be simply too bland—if eaten on the half shell—as compared to similar fare from the sea. The meats and the juices inside the shell of freshwater mussels are of course lacking in salt, iodine, and other briny ingredients.

(See also BIVALVED MOLLUSKS.)

N

narwhales (*Monodon monoceros*) A relatively small whale of the arctic seas, the narwhale has a body of about 15 feet, and the male sports an ivory tusk almost as long. Moreover, the tusk is twisted into a left-handed spiral. During the Middle Ages, the tusk was highly prized because it resembled that of the mythical unicorn and was believed to be of value in divining poison. (As late as 1789, instruments made of "unicorn horn" were used in France for testing royal food.)

The flesh of a narwhale is edible, but not highly regarded in most quarters. In Greenland, however, narwhale skin (with a layer of fat attached to it) is considered by the native Eskimo to be a delicacy and is eaten fresh and raw, frozen and raw, or preserved in blubber. Sometimes the skin is kept in caches for several years before it is eaten as *mattak*. According to Peter Freuchen (author of *Book of the Eskimos*), who partook of it, the skin doesn't become rancid during such long storage; instead, it ferments, and the skin takes on a walnutlike flavor. He also said that the blubber, green and sharp, tastes like roquefort cheese.

(See also WHALES.)

Nile perch (*Lates niloticus*) Anyone who considers the perch to be a small fish may be in for a surprise when angling in West Africa or on the Nile River and its tributaries. In these waters one can find the Nile perch, which weighs up to 300 pounds. Usually, larger nile perch are caught by using a smaller fish for bait. The tailwaters of the Aswan Dam are reported to be a good place to catch a big one. The fish is edible, but specimens that weigh over 20 pounds have coarse flesh.

The Nile perch is often confused with the smaller tilapia (*Tilapia nilotica* and related species), an edible African fish that has been stocked in Florida and is caught commercially and marketed as a food fish. Apparently, the name Nile perch sells better than tilapias. The tilapia has also been stocked in other places, from Syria to Ceylon, because it helps control the growth of algae, on which it feeds. Because of their feeding habits, tilapia are seldom caught on sportfishing lures, although a few are taken on earthworms. The tilapia weighs up to 5 pounds.

(See also FISH.)

noodlefish Sometimes called icefish or glassfish, these small cousins to the trout are transparent, slender, and fragile. About 14 species, such as the 4-inch *Salangichthys microdon*, live in Southeast Asia, inhabiting offshore salt water as well as estuaries, brackish water, and fresh water. Although the pencil-shaped noodlefish only reach a maximum length of 6 inches (depending on species), they are sometimes present in large numbers and are easily caught for food. They are considered to be delicacies in China as well as in Japan, where they are called *shirauwo*.

nutrias (*Myocastor coypus*) Sometimes called the South American beaver, the nutria is a large rodent, weighing up to 25 pounds, living in dens along streams, lakes, and canals, and feeding on vegetation on the shore. The nutria's reddish brown fur has commercial value, and many attempts have been made to raise it for market. Not many years ago, breeding stock was sold (at high prices) to people in the United States, France, and other countries. Most of the ventures didn't work out, and the animals were either released or escaped. In areas with suitable habitat and climate, they quickly became established in the wild. Today they are considered to be pests along the Gulf Coast of the United States. They can do considerable damage to dams, dikes, and agriculture. They also compete with native wildlife, such as muskrat, that depend on similar habitat; indeed, some southern states have liberal game laws and offer unrestricted trapping of the animals.

The nutria is edible, and the meat of the young animals has received high marks in some circles. Along the Gulf Coast of the United States, recipes for the nutria have even gained entry into some regional cookbooks.

(See also BEAVERS.)

O

octopuses and squid Although these mollusks have no external shells, they are related to oysters, clams, and snails. Both octopuses and squid are widely eaten in the Mediterranean area and in Asia, as well as other parts of the world. They are not, however, widely consumed in North America, although they are readily available in American waters.

Octopus

The common octopus, *Octopus vulgaris*, can be found in all the seas of the temperate and tropical zones. It was highly prized by the ancient Greeks and the Romans, and is still considered to be a delicacy by many peoples of the Mediterranean, as well as by the Chinese. Some species excrete an inky substance as an escape mechanism; this ink is sometimes used in recipes. Although the flesh can be quite tough (in Greece it is beaten over rocks before cooking), it is sweet and tender when properly prepared. According to *McClanes New Standard Fishing Encyclopedia*, "They are taken in pots, baskets, and traps, and by dynamite, spears, hooks, and a multitude of lures and poisons. They are eaten fresh, baked, fried, boiled, dried, canned, salted, and pickled. . . ."

There are a number of species of octopods, ranging in size from only 1½ inches long to the giant *Octopus dofleini*, which lives in deep waters near Alaska and which has an arm spread of up to 30 feet. The large octopuses can be a danger to man, but attacks are rare.

Squid is eaten a little more frequently in the United States than octopus, but it is still perhaps the most underutilized of all seafoods. Squid are, in fact, one of the most numerous animals in the sea, and they are very important as food for fish. They are eaten in most parts of the world, and are caught in seines, trawls, and traps. They are eaten fresh, salted, canned, or smoked. The Chinese are fond of dried squid, which are easy to market.

As a rule, squid are not the target of sport fishing, but the giant squid, or jibia (*Dosidicus gigas*)—which grow up to 6 feet in length—are sometimes caught by anglers off the coast of Chile. These are edible, but tough. As a rule, the smaller squid are better, but all of them tend to be tough if not prepared correctly. The key is to cook them for a very short period of time in high heat or for a long period of time in low heat.

(See also BIVALVED MOLLUSKS.)

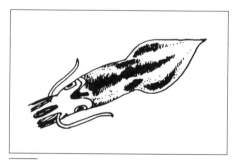

Squid

oilfish (*Ruvettus pretiosus*) According to *McClane's New Standard Fishing Encyclopedia*, the oilfish lives all around the world in the tropical and temperate regions of the oceans. It grows up to 100 pounds and, since it is usually stays in deep water (600 feet or more), it is not often taken by sport fishermen. It is kin to the mackerel, which has a general reputation for being quite oily but tasty. The oilfish, however, may cause severe poisoning if not prepared properly. The fish is nonetheless eaten in Polynesia, where it is boiled and the oil poured off. *McClane's* recommends that the fish not be eaten at all.

(See also FISH.)

okapis (*Okapia johnstoni*) Hearing rumors of a horselike creature that lived in the forest of the Belgian Congo, British explorer Sir Harry Johnston discovered, in 1901, what the pygmies called the okapi. It is indeed a large animal, weighing up to 550 pounds, but it turned out to be more akin to the giraffe (family Giraffidae) than to the horse. It is, however, much shorter of leg than the giraffe and doesn't have as much neck. Its reach into the trees for forage is less than half that of the giraffe. It feeds mostly at night, and often stays along the edge of streams. In color pattern, it isn't a jigsaw like a giraffe or striped like the zebra. Its body is a solid chestnut color, but its legs have white bands.

The okapi still can be found wild in Africa, but its range is restricted to a rain-forest portion of northern Zaire. Having such a limited habitat, the okapi is now protected by the government. On the other hand, the meat is so highly prized that sustenance hunting as well as commercial poaching is difficult to control. The animals have been kept in zoos and have bred successfully in captivity.

opossums Of the 25 species of opossums that live in the Americas, the Virginia species (*Didelphis virginana*) is the largest, and is one of the few mammals in North America that is expanding its range and multiplying its numbers. When the white man arrived in America, opossums didn't live much farther north than Maryland. Now they have expanded their range as far north as eastern Canada. They have also been introduced in California, and have become established in western Canada. In time they will no doubt fill the area in between. The animals have also been introduced to New Zealand. Part of the reason for the opossum's success is that it does very well on farms, in towns, and in inner cities.

The Virginia opossum weighs up to 12 pounds, and it is frequently eaten. In fact, the meat is highly praised by some people, who like it baked with sweet potatoes. At one time, opossum hunting was a rather popular sport, practiced mostly in the southern part of the United States. The hunting was done at night with the aid of a dog or two. The opossum was often taken alive, which was surprisingly easy to do since it feigns death when in danger. (Hence, the term "playing 'possum.") Since opossums are scavengers, many people in the South had special cages in which the animals were kept for several days in order to clean them out before they were eaten.

Even today, some annual events in small towns have an opossum theme, at which plates of "possum and taters" are sold for one worthy cause or other. There is also a national opossum club, which encourages people to eat more opossum and sells a bumper sticker touting just that.

The opossum is one of the oldest surviving mammals, feeding on dinosaur eggs 70 million years ago. In fact, it has been suggested that the lowly opossum might have been at least partly responsible for the disappearance of the dinosaurs. The opossums are marsupials and carry their young in a pouch like the kangaroo. The Latin name *Didelphis* means double womb, and the male has a forked penis.

Although the Virginia opossum is doing well and expanding its range, most of the other 25 species face a declining habitat. Many of them depend on tropical forests, which are fast disappearing in Central and South America. All of these species are edible and some are as large as house cats, but others (the mouse opossums) are only 2 or 3 inches long and are too small to be of much food value. (See also MARSUPIALS.)

orangutans (*Pongo pygmaeus*) Long-armed apes, orangutans are the largest arboreal mammal, with males weighing up to 200 pounds. One of the better-known apes, the orangutan is also known as "man of the woods." The name *ôrang ûtan* is from Malay, meaning jungle man. Today, the animal is called *maias* by some of the natives. These days orangutans live wild only in the remote lowland jungles of Borneo and north Sumatra, where they feed mostly on fruits. They will also eat eggs and insects, as well as tree bark and young plant shoots. But fruit is the main part of their diet, and they often sit all day in a single tree and gorge themselves. Water is usually dipped by hand out of tree holes.

Orangutans use their weight to bend limbs over, thereby going from one tree to another. They also swing on vines, as in the Hollywood version of Tarzan. The animals are highly intelligent and show remarkable memory and reasoning concerning the location of certain foods at certain times of the year. Sometimes they travel and feed on the ground, bending over bushes and small trees to get at tasty fruit such as lychees and mangoes.

Orangutans are threatened by the destruction of the forests in their homelands for timber and agricultural expansion. They have also been exported in considerable numbers because of their popularity in zoos. But, fortunately, they are now protected in several parks and preserves, where major populations still exist in the wild. Some natives in Borneo believe that the orangutan is the descendant of a man who fled into the forest after being disgraced by his tribe. The animal has a long call, lasting up to a minute, which some natives believe to be an expression of the loss of its human bride. In fact, some people even believe that the animals abduct human girls from time to time.

The orangutan has been eaten for food in times past, and may still be poached for food in the remote part of its range. At one time, the animal also lived on the mainland of Asia. In Vietnam, the lip of orangutan was once considered a delicacy.

ortolans (*Emberiza hortulana*) This small bird, a bunting, is highly prized as food in some areas. The species is quite common in Europe (especially in France, Spain, Italy, and Greece) as well as in parts of western Asia. Although not considered a game bird, the ortolan is netted in large numbers for food. In France, it

The Ostrich in Modern America

The U.S. herd, which has grown rapidly in the past several years to more than 10,000 birds, is found in states as climatically diverse as Minnesota and Texas. An average bird will yield a hundred pounds of meat that looks and tastes like beef but is lower in fat and cholesterol than chicken. By-products include boot leather and feathers for professional-quality dusters. The fledging U.S. ostrich ranches won't begin to slaughter their birds until they have much larger stocks, in five years or so. Ostrich meat is eaten as commonly as veal in many Western European countries. The world's supply has thus far come primarily from South Africa, which has managed to keep the bulk of its ostrich-raising information top secret (the most recent genetic study was published in 1961.) But zookeepers' experience suggests that a hen is capable of producing 50 eggs a year for more than 40 years—which suggests why at the moment a pair of ostriches can cost up to $80,000.

—*The Atlantic,* April 1992

is captured alive and fattened in cages for the table. Indeed, some French epicures insist that the ortolan must be cooked in its own fat.

The birds are sometimes wrapped in grape leaves and roasted. When well fattened, ortolans are also good for grilling over glowing embers or charcoal. The peasants of feudal Europe no doubt enjoyed them cooked before the open hearth. (See also BIRDS.)

ostriches (*Struthio camelus*) The largest extant bird, weighing up to 300 pounds, the ostrich makes excellent eating and was highly praised by the Roman epicures. Ostrich eggs are also edible, and were enjoyed by the ancient Egyptians and other Africans. The eggs weigh about 3 pounds each, and the hollow shells have been used to store water.

The ostrich lives wild in Africa—especially on the plains—where it has been widely hunted for its feathers as well as for meat and leather. The bird has keen eyesight, is wary and difficult to stalk, and runs at speeds of up to 40 miles per hour—all of which made it a difficult prey for bushmen armed with primitive weapons. In order to get closer to the birds, the primitive hunter wove a sort of movable blind with grass and stuck feathers into it. Wearing this on his back, he walked in a stooped position. He also selected a stick with a knot or bend in the end, about the size and length of an ostrich neck and head. Thus rigged, the hunter bent over, held the stick up, and walked closer and closer to a real bird. Sometimes, however, the ostrich took issue with the impostor and chased the hunter off.

Modern firearms and the loss of habitat have greatly reduced the numbers of wild ostrich. The birds are now raised commercially in Africa, France, the United States, and other countries for their feathers, leather, and meat. Whether they will ever become important to modern man as a source of food remains to be seen.

otters One or another of 12 species of otter live on land or in the water near land in most subpolar regions of the world, except for Australia and Madagascar. All are edible—but none is noted as table fare. The flesh of the otter is said to be oily and leathery. The animals were, however, eaten by ancient man and by some American Indians. According to Henry David Thoreau, the Indians of New England made a soup from lotus roots and fat from the North American river otter, *Lutra canadensis.*

oysters The newly hatched young of oysters, often called sprats, fix themselves to pilings and other objects in the water, where they remain for the rest of their lives. By comparison, other bivalved mollusks are free to move about, however clumsily. Indeed, the foot of the young oyster drops off and the adult oysters simply can't move about in the water. Oysters grow in the tropical and temperate regions of the world's seas, between latitudes 64 N and 44 S.

Oysters were widely eaten by primitive peoples of coastal regions, and they remain a popular food with modern man. Because they can be transported alive without being in water, they have also been enjoyed by inland connoisseurs. This is made possible because the oyster shuts its shells tightly when it is removed from the water. This seals in some salt water and the oyster carries part of its environment with it. When the oyster is thus sealed up and kept at a low temperatures, its heartbeat slows down and it requires little oxygen. The oyster's ability to live for relatively long periods of time enabled Abraham Lincoln to serve fresh oysters at parties in Springfield, Illinois.

Often oysters are eaten on the half shell. That is, the live oyster is shucked

and put back into the larger of its two shells. Then it is eaten raw, making it one of the few animals that is consumed alive by modern man. Fresh shucked oysters are also marketed in refrigerated containers, but these are thoroughly washed, robbing them of their salty taste and flavor. Of course, oysters are also canned and smoked for the market. In Asia, they are even sold in dried form.

The ancient Greeks were fond of oysters, and Aristotle noticed that the sprats would attatch themselves to objects in the water. This observation sparked the idea of putting suitable frames into the water to help give them a suitable platform—which is still the basis for modern oyster farming. The ancient Chinese also farmed oysters, as did the Romans. Today, oysters are farmed on a large scale in the United States, Japan, Holland, and other countries. The problem with modern oyster farming, as well as foraging, is in keeping the water free of contaminants. As a rule, oysters do best in saltwater bays that are fed by freshwater streams. Since they feed by filtering the water, they are susceptible to contaminants that may have entered the water far upstream. Indeed, one future use of oysters may not be for food, but rather to clean up the waters in certain areas.

To be sure, oysters from the wild are still eaten from various waters around the world. To a large extent, an oyster will take on its physical characteristics from its environment. An oyster found in deep water, for example, will look different from one discovered in shallow water. An oyster and its shell taken from the mouth of a dark river will look darker than one from crystal clear water. In America, local peoples all along the Atlantic and Gulf coasts defend local favorites. Yet, they are all the same oyster: *Crassostrea virginica*. Similarly, in Europe, the local favorites are all *Ostrea edulis*.

Some local favorites, however, are separate species. The mangrove oysters, for example, are highly popular in Trinidad, where they are sold from pushcarts or stands and are believed to have outstanding aphrodisiac powers as well as good flavor. These small oysters (*Crassostrea frons*) grow on the roots of mangrove trees, which stick down into the water like many fingers. In short, these roots make good frames for oysters. The problem is that at low tide the roots are out of the water. This forces the oyster to live out of the water and, of course, this species has developed to suit the environment. Along the Gulf and South Atlantic coasts of Florida (and on up into South Carolina) the mangrove oyster, being small, is not widely enjoyed except by racoons, who wade out at low tide and pick them off the mangrove roots. In fact, *Crassostrea frons* is often called coon oyster in the southeastern United States.

This rather strange oyster caught the attention of Sir Walter Raleigh during a trip to Trinidad. He ate one and pronounced it to be very salty and good. But he was laughed at when he reported to Queen Elizabeth's court that he had discovered in the New World an oyster that grew on trees.

(See also BIVALVED MOLLUSKS.)

P

paddlefish (*Polyodon spathula*)   Often called spoonbilled catfish or shovelnosed catfish, this large freshwater species isn't even related to the catfish. Its only close kin is a large paddlefish that lives in the Yangtze River of China. The following report on the American species comes from *Outdoor Highlights*, July 1991, published by the Illinois Department of Conservation:

> The American paddlefish, or spoonbill, is one of North America's largest, most unusual, and primitive freshwater fish: a holdover from the Stone Age that has changed little during the ensuing eons. It is found only in the Mississippi River and its tributaries, which extend from New York to the Dakotas and southward to the Gulf states. Though known in many areas as the spoonbill catfish or boneless catfish, it is not of the catfish family.
>
> Once numerous enough to be harvested everywhere in its range for food and roe, from which caviar is made, the paddlefish has undergone a drastic population decline since the turn of the century because of destruction of the fish's habitat and construction of dams that interfere with its spawning migrations.

Illinois is one of several areas where the species is not in trouble, however; its numbers are high enough statewide to support both commercial and sport fishing. . . .

Strangely, these large fish, which grow up to 7 feet in length and weigh up to 180 pounds, feed only on zooplankton, microcrustaceans, and tiny insect larvae. Consequently, paddlefish are not often caught by hook and line, although some are taken by snatching with treble hooks, especially in the swift tailwaters below dams on the Tennessee, Missouri, and other large rivers of mid-America. They are boneless fish and, when smoked or grilled, their sweet, oily flesh is highly regarded.

(See also FISH.)

palolo worms (*Palola viridis*)   These saltwater worms live around the coral reefs in the South Pacific. Once a year—in November at the third quarter of the moon—millions of the worms cast off their egg-bearing ends, which come to the surface and are said to "swarm." A smaller swarm may occur in October, but November is clearly the main event (called *palolo levu* in Fiji). It is the optimum time for harvesting the swarm for food, which is sometimes called Samoan caviar.

(See also EGGS.)

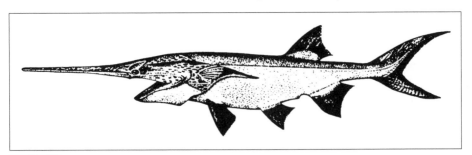

Paddlefish

pangolins (family Manidae, genus *Manis*, order Pholidata) In the past, there has been some confusion about the scientific classification of these animals. But anyone who encounters a pangolin in the wild won't ever confuse it with anything else—it looks like a heavily armored anteater. Tough, horny scales cover the entire body, except the undersides. In appearance, is has been compared to an artichoke. When threatened, the animal balls up like the three-banded armadillo, and is virtually impregnable to most predators. The name pangolin comes from the Malayan, meaning roller. They range in size from 3 or 4 pounds to 73 pounds for the giant pangolin (*Manis gigantea*). In all, seven species live in Africa and Southeast Asia.

All of the pangolins prefer to feed on ants and termites. Some live and hunt in trees, and others go after terrestrial insects. All species have curved claws, which aid in digging into insect nests. The tree pangolins sometimes feed on columns of insects moving along a tree limb, but they also use their claws to aid in climbing and clinging to limbs. In the terrestrial species, the claws hinder walking and are usually tucked in or balled up, which gives the animals a peculiar gait. When in danger, they can, however, run on their hind legs, using their tail for balance.

The pangolin is hunted in Africa for its meat and its scales. In Asia, the pangolin is hunted primarily for its scales, which are powdered and used in medicines as well as in various aphrodisiacs.

peacocks (*Pavo cristatus*) Technically, the headword for this entry ought to be peafowl, simply because peacocks are the males of the species. The females are peahens, which do not sport the spectacular tail of the male birds. Together, they are known as peafowl. By whatever name, they are the largest of the pheasant family and are native to India and Ceylon.

Easily tamed, the peacock was known to early traders from the West, and was even praised by the pharaohs of Egypt. Alexander the Great introduced the bird to Greece, from where its fame spread across Europe. India exported the peacock's feathers in large numbers. Soon the birds were raised in Rome as table fare for the gourmets and gourmands. Later, the common peoples of medieval Europe also raised the peacock, and it became popular as table fare, especially on festive occasions. Charlemagne, for example, served thousands of peacocks at a single banquet. Often the bird was served with a good deal of pomp.

But the peacock's culinary importance came to an abrupt end. After its discovery and domestication, the American wild turkey took over the strutting in the barnyards of Europe. Today the peacock is raised mostly for show and it is, to be sure, quite an attraction at the world's zoos.

(See also BIRDS.)

pea crabs These small crabs belong to the family Pinnotheridae, and the females live in oysters and other bivalved mollusks. They feed on the food that collects around the oysters' gills, and they don't eat or otherwise harm the oysters. Anyone who shucks enough live oysters will come across a pea crab and, of course, it will usually be alive inside the oyster's shell.

In some coastal areas, such as the Outer Banks of North Carolina, the pea crab is a delicacy. They can be steamed and eaten, but usually they are eaten alive as soon as they are discovered as a sort of oystershucker's bonus. When mature, they are pink, have soft shells and are up to an inch in diameter, although most individuals are much smaller. Being tiny, the crabs have no commercial value.

(See also CRABS; CRUSTACEANS.)

mals tend to live in roaming herds of 10 to 15, and they can do considerable damage to small fields before moving on.

The peccary is smaller than the wild boar and isn't as dangerous to hunt. The peccary does have tusks, but they point downward and don't slash as badly as the wild boar's.

(See also PIGS.)

Peccary

peccaries Three species of these piglike animals grow in America, from south Texas to northern Argentina. The collared peccary (*Tayassu angulatus*), the more widespread of the three, is the species that roams the wilds of the southwest United States, where it is called javelina.

The peccary is hunted for food and sport throughout its range. The meat receives high marks from some authorities, but not by others. A good deal depends on proper handling after the kill. In some areas, the peccary is considered to be a pest because it eats crops, especially corn and watermelons. The ani-

periwinkles These small marine snails with a spiral turban-shaped shell usually live between the high and low tide marks of rocks, pilings, and so on. In the tropics, some species affix themselves to mangrove roots. The largest of the species, which grows up to an inch and a half wide, is the common periwinkle, or winkle, *Littorina littorea*. It is common on the shores of northern Europe and northeastern North America. (Originally confined to the eastern part of the North Atlantic, *Littorina littorea* was planted in Halifax in 1858 and has spread southward.)

At one time, the winkle was an important source of food in the British Isles, and in France, especially among the French poorer classes, and winkle shops were popular in London. They were also called buckies in Scotland and willocks in Ireland. The winkle has been cultivated in France, where it is called *vignette, vignot, brelin, guignette,* or *escargot de mer,* depending on location. For many years the periwinkles were food for the French poorer classes, but these days they are considered to be a delicacy, although they are still sold by street vendors in some European towns along the coast.

Periwinkles are boiled for a few minutes in salty water before they are eaten. This loosens the snail from its shell and helps open the cap. Then the meat is pulled out with some instrument like a nutpick. Usually, it is eaten with a butter sauce. In Ireland, periwinkles are sometimes dipped into a fine oatmeal before they are eaten. They are also used in a soup, along with carrageen sea moss and milk.

(See also SNAILS.)

petrels These seabirds, a diverse group comprising many species, range far and wide around the globe, and may have a distribution greater than any other group of birds. The curious diving petrels frequent the South Atlantic, and they seem to fly under the water, breaking the surface at full speed. Many of them even appear to walk on water.

The petrels are no doubt eaten from time to time throughout their range, but they are not significant as a regular food supply except at some island nesting areas. The Tristan de Cunha group (three small and remote volcanic islands in the South Atlantic) hosts millions of petrels during nesting season; the birds and their eggs have been taken from there in large numbers by settlers and by seafarers.

(See also BIRDS.)

pigeons and doves Almost 300 species of these birds, belonging to the family Columbidae, live throughout the tropic and temperate zones of the world. They are widely hunted for food and sport, and pigeons are raised for food. Usually, only the very young domesticated pigeons are eaten, and these are usually called squab or sometimes *pigeonneau* on French menus.

Domestic pigeons have been traced back to 3000 B.C. in Egypt. The practice of using homing pigeons as messengers goes back at least to Genghis Khan, who used them during military expansion. Of course, messenger pigeons have been used in modern times; in one instance a bird in the service of the United States Signal Corps flew a record distance of 2,300 miles.

At the turn of the 20th century, some 9 billion passenger pigeons lived in North America. Over the next few decades, these were killed extensively by market hunters, who apparently kept the bird from breeding. The species is now extinct. These days, the mourning dove (*Zenaidura macroura*) is by far the most plentiful and popular game bird in North America, and it makes excellent eating. Its feeding habits—flocking from one field to another, wherever the pickings are better—allow it to make maximum use of the grain wasted by mechanical corn pullers and other modern harvesting devices.

The white wing dove and other species are hunted in Mexico and South

America, and these are sometimes present in large numbers. Pigeons are also hunted in Egypt and other places. Scientifically, there is no distinction between a dove and a pigeon but, generally, a dove is considered to be smaller than a pigeon. The largest birds of the group are the crowned pigeons (genus *Goura*) of New Zealand, which grow as big as chickens and are hunted there for food and sport. (See also BIRDS.)

pigs Wild pigs and boars, as well as feral pigs, are eaten all around the world. The wild boar or Eurasian wild pig (*Sus salvanois*) of Europe, Asia, Sumatra, Japan, Taiwan, and North Africa has been widely hunted in the past and is a dangerous game animal. It's eaten by sportsmen, and is sold commercially and in restaurants. It has been introduced into some countries, including those of North America, usually to highly managed hunting preserves.

Eight other species of wild pig or boar live in various parts of the world, including the giant forest hog (*Hylochoerus mienertzhageni*) that lives in the Congo basin and other parts of Africa. This animal grows up to 600 pounds in weight.

Domestic pigs are used for food in most parts of the world, and are popular in countries such as China. In fact, pork is the world's most popular meat—in spite of the fact that it is forbidden food in two major religions: Judaism and Islam. One reason for the high demand for pork is that it can be cooked and eaten in a number of different ways. Another is that the pig grows fast and is efficient in converting what it eats into meat. In general, pork is less expensive than beef or lamb. The meat is sold fresh as well as frozen and cured. Before mechanical refrigeration was invented, salt pork was a very important meat.

At one time, pigs were kept in towns and cities, where they cleaned up the garbage. Country people allowed their pigs to forage, sometimes without fences to keep them in. On some of the islands of the Outer Banks of North Carolina, for instance, fences were put around the houses instead of around pastures. Hogs can, in fact, fend quite well for themselves in many places and can be raised on beech mast, acorns, chufas, and other wild foods, including worms and snakes. They are great rooters. Such free-roaming hogs, tend to be rather wild, however, and not many years ago many farmers had "catch dogs" to help in the roundup. These days, however, most of the 400 million domestic pigs aren't allowed to roam through town and country. They are carefully raised and fed in controlled environments.

In addition to pork, the pig provides fat (lard), pigskin for gloves and footballs, and casing for sausage. Its hair is used for upholstery and insulation. Other American foods produced from pig meat and by-products include ham, bacon, mountain oysters (pig testicles), pig's knuckles (boiled and pickled), chitterlings (fried intestines), pig's feet, and scrapple (mashed patties of pig leftovers).

Nobody knows for sure, but some scholars believe that the pig was first domesticated in China, probably from *Sus salvanois* or one of its cousins, and that its original use was primarily as a scavenger. In any case, a number of different varieties of pig have been developed in domestication, including such fine pork producers as Duroc,

Poland China (developed in Ohio), Tamworth, Palouse, and so on.

(See also BABIRUSA; PECCARIES; WART-HOGS.)

piranhas (family Characidae) Once defined as fish that are eaten by man and vice versa, piranhas are native to South America. Growing up to 2 feet long, they are often caught by local anglers using cut bait or strip bait and wire leaders; from time to time they will hit artificial lures. They make excellent eating.

Although dangerous, piranhas seldom attack man or other large animals. They do, however, feed voraciously when they are present in schools of 30 or more. They have been known to clean a 100-pound capybara—a large edible rodent of South America—to the bone within a minute. Apparently they are attracted to blood—human or otherwise. Piranhas have exceedingly sharp, pointed teeth that mesh together. These teeth have been used as arrowheads by some natives of the Amazon Basin.

Curiously, the wimple-piranha feeds on the scales of other fish. Their lower jaws are longer than their upper jaws, which enables them to scrape the scales off another fish with an upward bite.

(See also FISH.)

pirarra (*Phractocephalus hemiliopterus*) The great Amazon Basin in South America holds some 2,000 species of fish, including several very large catfish. In an article about these catfish published in the *1989 World Record Game Fishes*, Gilberto Fernandes said, "In our Indian language the word *pira* means fish, and *arra* means a big and colorful parrot with long tail feathers." A Brazilian, Fernandes holds the all-tackle record for

pirarra—90 pounds 6 ounces. Bigger catfish grow in the Amazon, including the piraiba (*Brachyplatystoma filamentosum*). Officially, Fernandes also holds the all-tackle sportfishing record of 256 pounds 9 ounces for piraiba, but larger ones have been taken in nets and on trotlines. Reports of catches up to 500 pounds have been made in the past.

Some of these large catfish make very good eating, and the meat is exported to other parts of South and North America. Although eaten, the pirarra is not highly prized by the natives in the area, partly because the meat is as red as beef, whereas the quality of most catfish is judged in part by the whiteness of its meat. The roe of pirarra, on the other hand, is a delicacy. The fish is very fat, and Fernandes says that the Indians use its fat to feed parrots, which causes the birds to turn from green to yellow.

(See also CATFISH; FISH.)

pirarucús (*Arapaima gigas*) These freshwater giants, growing up to 15 feet long and weighing in at over 500 pounds, live in tropical South America. They are ugly creatures, resembling the prehistoric bowfins of the United States. Their scales are large and thick, and their tongues—said to be used as files—are bony. The pirarucú will hit artificial lures, especially plugs, as well as natural bait, and is sometimes taken on rod and reel by sport fishermen. But most of the large ones are taken by the natives in nets and on trotlines. Because the fish has a habit of lazing and rolling near the surface, they are also hunted with harpoons and arrows. In recent years, the fish seems to have declined in numbers, but it is still plentiful in the Amazon and in the tropical areas of some of its tributaries. It is also found in lakes and in the large streams of the Guianas.

Firm and white, the flesh is highly prized by the natives. It is sold both fresh and salted in local fish markets, where it brings top prices. The fishery is of some commercial importance, especially in Brazil. Some of the "commercial" fishing is not done by large companies, but by individuals fishing or hunting from small boats. Some natives even run trotlines for large fish by wading. When they catch a 200-pounder, they somehow subdue it and haul it to the bank over their shoulders.

(See also FISH.)

platypuses (*Ornithorhynchus anatinus*) Often called duckbilled platypuses, or simply duckbills, these strange creatures live only in southeastern Australia and Tasmania. When the first specimen—actually, a dry skin—arrived in England around 1798, it was considered to be a joke or at least a fake. A taxidermist, it was believed, had attached the bill of a large duck to a mammal. It turned out to be real—but was it a mammal? The thing laid eggs like a duck or a reptile. Some people even speculated that it was a reptile with fur.

The platypus has webbed feet and a large, flat tail. In short, it looks like a small beaver with a pliable, leathery bill and a paddle tail with fur on it. Adults weigh up to 5 pounds, although weights vary widely among individuals and with seasons. The platypus lives in dens under the banks of streams and lakes. It feeds entirely on aquatic invertebrates, mostly on those that live on the bottom of the stream or lake. The duck bill no doubt comes in handy for rooting up these creatures.

At one time, the platypus was hunted for food by the original Australians and early settlers and became nearly extinct. Fortunately, it became a protected species and its populations have increased in some areas during the 20th century.

porcupines Rodents covered with sharp spines, or quills, porcupines are classified in two families. The New World porcupines include 11 species in the family Erethizontidae. The North American porcupine, *Erethizon dorsatum*, lives in most of Canada and the United States, except for the southeastern states. The Old World porcupines, family Hystricidae, include 11 species that live in Africa, southern Europe, and Asia.

Both kinds of porcupine are eaten, especially in Africa, where the animals are often a menace to agriculture. Generally the flesh of the porcupine, like that of the opposum, is rather fatty. The liver of the North American porcupine is said to be gourmet fare.

(See also RODENTS.)

porpoises See DOLPHINS AND PORPOISES.

prairie chickens (*Tympanuchus cupido*) This North American grouse inhabits the planes of the Mississippi Valley from Louisiana to Manitoba. At one time, they were a popular game bird, and they are still legally hunted for sport and food in some states. In other states, they aren't faring too well. Here's a report from the Prairie State, reprinted in part from *Outdoor Highlights*, published by the Illinois Department of Conservation on June 17, 1991:

> When European settlers arrived in Illinois, prairie chickens thrived on the native tallgrass prairie that blanketed two-thirds of the state. Ironically, the

clearing of forests for agriculture initially was a boon for this native grouse. With more clearings for them to inhabit, prairie chicken numbers expanded and millions of the grayish-tan birds with the dark brown horizontal banding on their feathers boomed and whirred on the prairie. Nineteenth century market hunter H. Clay Merritt, for example, writes of kicking up a prairie chicken flock from a 40-acre field containing more birds "than all the cities in the Union could consume in a month."

Unregulated market hunting played a role in the prairie chicken's demise in Illinois. The fact that Chicago markets measured prairie chickens they bought in "tons" or "cords" is testimony to the volume of prairie chickens being harvested in the mid 1800's by market hunters, and Capt. Adam H. Bogardus' claim of shooting 600 prairie chickens in 10 days may not be unusual. But the damage done to the prairie chicken by the market hunters' guns was nothing compared to the devastation wrought by the invention of the self-scouring plow.

Subspecies also live in Kansas and west Texas, and a coastal species (*Tympanuchus attwateria*) lives in coastal Texas and Louisiana.

prairie dogs Several species of these rodents of the squirrel family live in North America. The most important of these—the black-tailed prairie dog (*Cynomys ludovicianus*)—lives in interconnected burrows on the low plains as well as on the plateaus of the Rockies, ranging from southern Saskatchewan to northern Mexico. At one time, the rodents existed in large "towns," which sometimes covered 160 acres and contained thousands of prairie dog residents.

Prairie dogs were eaten by the American Indians as well as by some settlers and modern-day varmint hunters. The Indians sometimes cooked them by merely tossing them into the embers of a campfire until they popped open. Kit Carson, the celebrated plainsman, is said to have enjoyed prairie dogs grilled over a campfire.

Prairie dogs have been vastly reduced in numbers, mostly because they were considered pests. Their reduction has also caused a shortage of burrows, which are also used by owls, rattlesnakes, foxes, and other critters.

(See also RODENTS.)

prawns See SHRIMPS AND PRAWNS.

pronghorns (family Bovidae, genus *Antilocapra americana*) Often called pronghorn antelope, these animals are really members of the family Bovidae. In fact, they are the sole existing members of the subfamily Antilocapridae, and are classified somewhere between deer and cattle. The animals are also called prongbuck. In Mexico, they are known as *berrendos*, and there are some local Indian names as well. By whatever name, the pronghorn is excellent eating if it is properly handled, and if the contents of the scent glands are kept off the butchered meat. Adults weigh about 150 pounds.

In modern times, pronghorns have lived on the plains of the western part of North America, ranging from Canada to upper Mexico. The animals take long strides—up to 27 feet—and can run up to 55 miles per hour. They are highly regarded as game animals by hunters. Although their numbers were reduced drastically—from an estimated 50 million in 1850 to 13,000 by 1920—hunting

Pronghorn

regulation and management have helped the animals make a comeback.

The pronghorn has a rather pronounced patch of white hair on its rump. This patch of hair can be erected and thereby serves as a form of communication between pronghorns. (The pronghorn also has rather strong scent glands located in the rump area, which are also used to signal alarm.) Although they are quite wary, pronghorns are also curious and have a habit of investigating patches of white similar to those on their own rumps. This habit enabled the early hunters to lure the pronghorn into gun range with white handkerchiefs tied to poles. Called flagging, this method of hunting is now illegal.

puffers When disturbed, these fish with highly poisonous innards inflate themselves with either air or water to an almost spherical shape. A number of species live around the world in tropical and temperate waters. They are eaten by local anglers in some areas and are even sold commercially. In America, the northern puffer (*Sphaeroides maculatus*) is sometimes marketed under the name sea squab. It has firm, white flesh of mild flavor.

In some puffers, poison is secreted by the innards, all of which should be discarded, including the roe. Some experts recommend that only the lower half of some species be eaten. In any case, the puffer has a loose skin, which slips off very easily, so that the fish is easy to clean.

In Japan, the puffer is considered to be a delicacy. According to *The Encyclopedia of Aquatic Life*, even the poisonous parts are eaten—after being prepared by a trained chef—under the name *fugu*. At one time, the Japanese used the innards of the puffer to commit suicide. In Hawaii, the word for the puffer fish is *makimaki*, meaning deadly death. Years ago warriors there used the gall to poison their spearheads. Generally, puffers are more toxic in spring or summer just before spawning. Also, puffers in tropical waters are believed to be more dangerous than their northern cousins; some authorities recommend that they not be used as food.

The puffer has a number of local names, including blowfish, swellfish, globefish, rabbitfish, and sea squab. Some species have spikes on their skin that stick out defensively when they blow themselves up, while others are more smooth-skinned.

R

rabbits and hares All together, some 44 species of rabbits and hares live in many parts of the world. Often, however, it's difficult to tell which is which. The well known long-eared black-tailed jackrabbit (*Lepus californicus*) of the American West is really a hare, and the Belgium hare is really a domestic variety of the European rabbit (*Oryctolagus cuniculus*), which is especially noted for the quality of its meat. Even more confusion arises by calling other animals rabbits or hares; for example, the Patagonia hare is really an agouti, which is also edible.

Although hares tend to run larger than rabbits, size is not always a reliable guide. Their breeding and reproductive habits are what distinguish hares from rabbits. All hares are born in open nests above ground and at birth are fully haired, wide-eyed, and almost ready to hop about. Rabbits, on the other hand, are born in fur lined nests in holes in the ground, naked, shut-eyed, and very much in need of parental care.

Both rabbits and hares were eaten by primitive man, and their bones have been found in ancient kitchen middens. Strangely, the Greeks shunned rabbit, believing that eating them would cause insomnia. The ancient Romans, on the other hand, more or less domesticated rabbits and raised them in enclosed areas called *leporaria*. They considered roast rabbit to be good eating; but the real delicacy was the unborn or newly born rabbit fetus. In the dark ages of Europe, some monasteries also raised rabbits in special gardens and the fetuses, along with poultry eggs in various stages of embryonic development, were considered to be "not meat" by the ascetic monks. Perhaps these were the original "Easter eggs."

Rabbits are quite popular as small game, and they are often hunted with the aid of hounds. In Europe, ferrets are sometimes used to chase rabbits out of their holes in the ground. Sometimes rabbits are hunted by a group of people armed with clubs, usually by encircling a field and driving toward the center. Large nets have also been used to catch rabbits during drives. Usually, a community hunt or drive is staged not only to gain meat but also to thin down the rabbit population in a certain area. Large numbers of rabbits can do considerable damage in some agricultural settings.

In general, rabbit meat is of excellent quality—firm and low in fat—and some connoisseurs believe that wild rabbits are better than domesticated ones. In America, the various cottontails receive high culinary marks. The European rabbit is also highly regarded. This species is the granddaddy of about 50 domestic varieties, and it has been introduced in Australia and other countries. During the heyday of large sailing vessels, the European rabbit was even stocked on remote ocean islands to provide food for shipwrecked sailors.

raccoon The common raccoon (*Procyon lotor*) is widely distributed in Canada, the United States, Mexico, and Central America, and it has been introduced in Europe and Asia. The term "raccoon" comes from an Algonquian word *ärähkuneum*, meaning "he who scratches with his hands." Often called coon, raccoons are curious about everything, and, indeed, shiny objects such as pearl buttons are used to bait traps set for the animals. Raccoon make excellent pets, but they are rather devilish or destructive in their play. The coon is trapped for its pelt, and coon hunting is quite popular in some parts of North America. The animals stir about mostly at night, and the hunting is done with the aid of dogs, many of which are special breeds. (There is a coondog graveyard in northwest Alabama, in which only coonhounds are buried, complete with headstones and inscriptions.)

The raccoon is edible and can be very good eating if properly prepared. Unfortunately, chasing the raccoon for hours with dogs and then making it fight for its life is not the best way to obtain good meat. Even the best of prime beef would taste gamy if cattle were slaughtered in this manner. In some areas, raccoon are very plentiful and are considered to be pests. To a large extent they feed along the water's edge, eating crayfish and other water creatures, but they will also eat earthworms, fruits, and grain. They have a fondness for corn on the cob, and this gets them into trouble with farmers. In the southern part of the United States, the best raccoon for the table are those that grow fat on peanuts left in the fields after the harvest.

In addition to the common raccoon, five other species live in Central and South America and in the West Indies. Also, the raccoon family (Procyonidae) includes the coatis, the ringtail, the panda, and other species. All of these animals are edible, but being nocturnal, they have not been used to a large extent by hunting peoples as food. Of course, the panda (especially the bearlike giant panda) has a very limited range and was not "discovered" until 1937. Even in

their home range, pandas are more or less protected by tradition as well as by recent laws.

The ringtail, which grows in dry areas of the American West from Oregon to Mexico, has no doubt been eaten but is not regular fare. Sometimes called Miner's Cats, ringtails are said to have been been more or less tamed and used to catch rats in camp. They also caught and ate lizards and maybe snakes and scorpions, all of which were of concern to miners.

Cooper's Entrecôte

During the siege of Paris in 1870, the French coopers skinned rats, marinated them in a mixture of olive oil and chopped shallots, and then grilled them whole over fires made with the wood from broken-up wine barrels. Called *Cooper's Entrecôte*, it was a forerunner of other grilled French dishes.

rats The common black rat, *Rattus rattus*, which can also be brown in color, probably originated in Asia and reached Europe in the 13th century, where it quickly populated the towns and cities and helped spread the bubonic plague. From Europe it traveled to other parts of the world aboard ships. Another species, the Norway rat or brown rat, *Rattus norvegicus*, also originated in Asia. It reached Europe in the 18th century, and has displaced the black rat in some areas. Both of these rats have been eaten, along with many other species of mice and ratlike rodents. For the most part, however, rats are eaten in most lands only during times of famine and war.

Before mechanical refrigeration was invented, rats were often eaten aboard ships, and were purchased at exorbitant prices from the ship's ratcatcher. They have also been skinned and sold as black-market squirrels. Even today rats are popular in parts of China, and are even served in a Canton restaurant, which prepares them in 17 ways, including "golden rat." These delicacies are, the restaurateurs are quick to say, good farm rats, not sewer rats. Rats are also eaten as everyday food by some of the common peoples of China, where they are called household deer. As a rule, the rats that feed on grain—as many do—are in fact quite tasty.

G. Gordon Liddy said that he ate a rat—fried the American way. Also, Henry David Thoreau is reported to have said that he enjoyed fried rats, served with a little relish. But he may well have been talking about muskrats, which no doubt lived along Walden Pond.

There are many kinds of rats and mice in various parts of the world. In fact, over a quarter of all species of mammals are rats and ratlike rodents. All of them are no doubt edible, and a few are eaten on a regular basis.

(See also AFRICAN POUCHED RATS; DORMICE; GERBILS; HAMSTERS; MUSKRATS; RODENTS.)

rays and skates These flat fish have large pectoral fins, which are aptly called wings. They include over 300 species, most of which are easily recognized as raylike because of their pancake flatness and overall shape. The exceptions include the guitarfish and the sawfish. All of these fish are eaten in most parts of the world, but are not highly regarded in the

markets of North America. The wings of rays have often been stamped into rounds and sold as scallops, but the meat simply doesn't sell well under the name ray or skate. Tastes and customs change, however, and the skates and rays are increasingly recognized as very good eating.

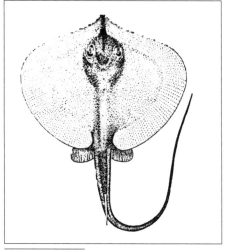

Atlantic stingray

Sportsmen often catch skates and rays, but as often as not they don't eat the fish, partly because they are scared off by the names stringray and electric ray, both of which refer to types of rays that can cause painful encounters. The electric ray can indeed generate up to 400 volts. The largest of the stingrays, the Indo-Pacific smooth stingray, weighs up to 750 pounds and can be dangerous to man. Reportedly, the poisonous spines at the base of the tail have caused the death of several humans near Australia. Two species of stingrays live in fresh water in South America. They tend to bury themselves in mud in shallow water, and they are feared by wading natives.

The largest of these fish are the Atlantic manta (*Manta birostris*) and the Pacific manta (*Manta hamiltoni*), which can have a wingspan of 20 feet and a weight of 2 tons. These giants are edible, but the smaller specimens are better table fare. The main danger from these large rays is that they sometimes overturn small fishing boats with their fins.

The sawfish, although flattened, is different from most rays in that it has a long bill with sawteeth. It is also more fish-shaped. The common sawfish (*Pristis pectinata*) lives on both sides of the Atlantic, usually along coastlines in tropical or temperate waters. It is also found in bays and freshwater rivers. These are large fish, and can grow to a length of 24 feet and weigh as much as 5,000 pounds. The saws of such fish have razor-sharp teeth and are of course very dangerous; although they don't attack man, they must be handled with care by fishermen. Sawfish are edible, and the smaller ones are considered to be excellent food. The sawfish is frequently eaten in the tropical parts of Africa and in Madagascar. (See also FISH.)

reindeer (*Rangifer tarandus*) Called caribou in America, these animals live along the fringes of the Arctic in North America, northern Europe, and Siberia. Excellent venison, they are raised commercially in Europe, and domesticated reindeer have been introduced to Canada and Alaska. Many other reindeer are herded, or managed, in a more or less wild state. In Lapland, some people depend on the reindeer for food, milk, and leather, and they also use them as draft, pack, and decoy animals. A few of the Lapps are still nomadic and follow the herds at least part of the year.

Some of the nomadic tribes of Siberia also depend on the reindeer, and they also follow the herd. They also use reindeer as draft animals, principally to pull sleds. All of the wild reindeer move about in herds, feeding on lichens or reindeer moss, which is also eaten to some extent by man. In America, inland Eskimo at one time depended heavily on the reindeer and also followed the herds for hundreds of miles. In addition to meat, the animal's hide was the preferred leather for use in clothing.

The early Eskimo usually hunted the reindeer with bow and arrow. When the animals were caught swimming rivers and lakes, however, the Eskimo usually used the spear. In spite of the fact that the word Eskimo is from a Cree Indian word meaning "eaters of raw meat," the Eskimo much preferred to eat cooked reindeer meat. In Eskimo cultures, the women usually cooked the meat, but they were forbidden to eat a great delicacy of the north country: boiled reindeer tongue. This dish (sometimes smoked) is also highly regarded in Scandinavia and Finland, along with reindeer bone marrow.

Although neither the domesticated nor wild reindeer are very important to modern man as a whole, it might well have been the very first mammal to have been domesticated. One theory has it that the reindeer, feeding mostly on mosses and ferns that grow on lands watered by melted snow—water that would be relatively mineral-free—suffered from a salt deficiency, which might explain their long trips to the sea or to inland salt licks. Even the salt in human urine (which would, one would assume, be deposited just outside the cave entrance) may have been enough to make some of the animals dependent on man to such a degree that domestication followed in due time.

remoras There are several species of remoras, such as the sharksucker (*Echeneis naucrates*), that have a suction disk on top of their heads. (They are often called suckerfish.) With the aid of this disk, they attach themselves to large fish and sea turtles, thereby getting a free ride. Believe it or not, primitive peoples in various parts of the world tied a line to the tail of the remora and used it to catch larger fish as well as turtles.

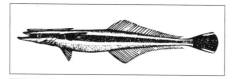

Remora

The use of the remora for fishing was reported in Europe in 1504 by a writer who said he got the information directly from Christopher Columbus, who in turn learned of the practice from the Arawak Indians of Cuba. The Arawaks kept a few remoras in pens at the edge of the sea and loosed them whenever a large turtle was sighted. Nobody believed the account until further reports came in from Japan, India, China, Venezuela, Australia, Malaya, Zanzibar, and elsewhere.

In any case, the remora itself is edible, and it was in fact customary for the aborigines of Australia to eat them along with the catch. In most cases, however, the remora was kept for future use in fishing.

reptiles See ALLIGATORS AND CROCODILES; SALAMANDERS; SEA SNAKES; SEA TURTLES; SNAKES; TORTOISES; TURTLES.

rheas These flightless birds live in South America. One of the three species, *Rhea americana*, ranges from Paraguay to Patagonia. Another species lives only in Patagonia, and a third lives only in northeast Brazil. Sometimes called the American ostrich, rheas have the same habit of feeding with grazing mammals, such as deer and guanacos. They were once important as food because of their large size, and they were often hunted with the aid of bolas, which tangled around their long legs. They are still eaten today and are relatively plentiful in Patagonia. In addition to their meat, rheas are highly prized for their feathers.

(See also BIRDS.)

rhinoceroses Five extant species of these large, ill-tempered animals now make up the family Rhinocerotidae. The largest of these—the white rhinoceros (*Ceratotherium sinum*)—weighs up to 5,000 pounds, and among land mammals is second only to the elephant. The white and the black rhino (both of Africa) have two horns in tandem, with the larger one usually forward. The Sumatran rhino also has two horns, while the other two Asian species have only one.

Rhinoceros

One source, *Larousse Gastronomique*, rates rhino meat as only fair—better than the elephant but not as good as the hippopotamus. Nonetheless, the meat is highly prized in some areas and its thick skin is considered to be a delicacy in parts of Asia. In all probability, the animal has not been a target for meat hunters in Africa and Asia because it is surprisingly agile for its size and is prone to charge instead of retreating, as many jeep-riding photographers have found out. The rhino is also difficult to bring down, partly because of its bulk and partly because of its thick skin. It is, in short, a dangerous animal for the hunter to face, even with a modern rifle. It wasn't meat that brought about the rhino's decline, but rather the value of its horns and destruction of its habitat.

rice fish (*Monopterus albus*) Like other members of the swamp eel family, these freshwater fish can exist in poorly oxygenated conditions. They can, in fact, live for a considerable length of time if they are merely kept moist. This characteristic, together with their edible flesh, makes the species a valuable source of food in parts of Asia, where proper refrigeration is not always available.

The rice fish grows nicely in irrigation ditches and often feeds in flooded rice fields. They are sometimes called rice eels.

(See also FISH.)

rodents (order Rodentia) Nearly 40 percent of all mammals on Earth are rodents of one sort or another. All of these are edible; many have been eaten from time to time, and a few are eaten on a regular basis. Two rodents—dormice and guinea pigs—have even been raised and bred especially for meat.

Rodents include three major groups: (1) squirrels and squirrellike animals, including beavers and springhares; (2) rats and mice and micelike animals; (3) cavylike animals, which are especially plentiful in South America and include the guinea pigs, capybaras, chinchillas, coypus, porcupines, and so on. Rodents also include hamsters and gerbils, both of which are eaten. Rabbits are not considered to be rodents and belong to the order Lagomorphs. Lemmings and voles are also edible, and have no doubt been taken for food in large numbers during migrations, but they are not everyday fare.

(See also CAPYBARAS; CAVIES; CHINCHILLAS AND VISCACHAS; DORMICE; GERBILS; GUINEA PIGS; HAMSTERS; SQUIRRELS.)

S

salamanders These secretive reptiles usually live in damp places, and some are entirely aquatic. About 350 species of salamanders (and newts) live in the more temperate regions of the world, and a large percentage of them belong only to North America. All salamanders feed on animal life, including worms, insects, snails, and so forth. As a rule, they are nocturnal, hunting at night and hiding by day.

All salamanders are edible, although most are too small to be of much value to man except on a survival basis. Some, however, are surprisingly large. The Japanese giant salamander (*Andrias japonicus*) grows up to 5 feet long, and the Chinese giant salamander (*Andrias davidianus*) also reaches a considerable size. Neither of these species leave the water, and they are sometimes caught by fishermen with baited hook. They are considered to be excellent eating.

The American hellbender, which grows up to 28 inches long, is also edible and is sometimes taken by fishermen. In the southeastern part of the United States, the greater siren (*Siren lacertina*) and the two-toed amphiuma (*Amphiuma means*) are taken by bait fishermen (often when they are angling for catfish at night) and are sometimes mistaken for and eaten as eels. In fact, the two-toed amphiuma, which reaches lengths of up to 30 inches, is sometimes called conger eel or lamper eel and is quite tasty.

sand eels Several species of small elongated sea fish called sand eels or American sand lance live over sandy bottoms in shallow water. The sand eel has an eellike body with a forked tail. When frightened, and sometimes on low tide, the fish bury themselves head first into the sand, usually no more than 6 inches deep. They are often caught by foragers digging for clams.

One species, *Ammodytes americanus*, lives along the coasts of North America. Other species live in Europe and on the east coast of South Africa. In France sand eels are called *lançon*. Although the fish is not commercially important, it is highly regarded as food by those in the know. They are usually deep fried crisply, and are often regarded as whitebait.

(See also FISH; WHITEBAIT.)

scallops Most of these bivalved mollusks move about through the water by opening and closing their shells, which causes a jet action. In order to do this, they have a strong muscle that attaches the top shell to the bottom. This muscle, known as the eye, is the part that is usually eaten, especially in North America; the eye is cut from the shells and marketed fresh or frozen.

The eye of the scallop is usually sautéed or otherwise cooked. In Peru, however, the scallops are often eaten raw on the half shell—like oysters—and all of the animal is eaten, not just the eye. The French and other peoples also eat the whole scallop. Most of the commercial scallops' eyes are, by the way, soaked in water for several hours, which makes the meat whiter and increases the weight—but robs the scallop of some flavor. (The same can be said of bucket oysters that are washed before going to market.) In some places, the meat of sharks, rays, and other fish are cut into cylinders and sold as scallops in fish markets as well as in some restaurants.

In North America, the large deep sea scallop (*Placopecten magellanicus*), which is dredged off the coast of the northeastern United States and Canada, is of primary commercial importance. The smaller and better bay scallop is also marketed, and is often taken by foragers who wade about and net it on flats of suitable bays, which usually have some grass growing up from the bottom. All together, 400 species of scallops grow in the seas of the world, and some gain commercial importance in certain areas. Usually, the shallow-water species have been important mainly as a part of the diet of the local peoples.

Scallop shells have been widely used as utensils by primitive man, and have been a symbol in art and architecture as well as in mythology. The goddess Aphrodite (Venus), for example, was believed to have arisen from a scallop shell. (See also BIVALVED MOLLUSKS.)

scorpionfish (*Scorpaenidae*) Several species of these ugly fish live in salt water and some of them are highly prized as food in various places. Although some species grow to about 17 inches in length, there isn't much meat left when their large head is removed; thus quite a few scorpionfish are simply too small to be of much value as food.

Poison glands are located at the base of the dorsal fin spines, and this venom can be injected through the spine. Since the wounds are very painful, scorpionfish should be handled carefully. (See also FISH.)

sea cucumbers Although edible, these echinoderms (class Holothuroidea) take their name more from their shape than from their taste. In short, they look like a cucumber with whiskers. The body is oblong, and grows from an inch to several feet. There are about 500 species, which may be found in shallow or very deep water, but only a couple of dozen species are commonly eaten. Usually, they are gutted, parboiled, and smoked or dried. They are also pickled. Sea cucumbers are harvested commercially and processed in the Indo-Pacific reef area, and are shipped to China and other places, including Madagascar.

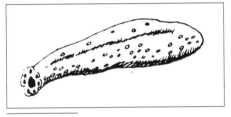
Sea cucumber

Dried sea cucumber is marketed in Japan as *iriko* and in the Philippines as *trepang*. They are also widely called *Bêche-de-mer*, from the Portuguese *bicho da mar*, meaning sea worm. (They are also called sea slugs.) A sea cucumber soup has been marketed as a gourmet item under the name *Bêche-de-mer*. It tastes somewhat like turtle soup.

seals (family Phocidae) A number of seals, sea lions and related mammals live in salt water and are usually associated with the Arctic or the Antarctic, although some species do venture far into warmer waters. A small 3-foot seal (*Pusa sibirica*) lives in fresh water, namely Lake Baikal in Siberia, but the rest of the family lives in salt water. Some, such as the 16-foot elephant seal of Antarctica, are huge. Most of the seals feed on fish, squid, and crabs, but one species—the leopard seal of the Antarctic—feeds primarily on penguins.

The Eskimo of the arctic regions used the seal and related sea mammals in a number of ways. The skin and fur were used for clothing, and the oil (rendered from the fat or blubber) provided fuel for heat and light. The Eskimo ate the flesh and were very fond of the so-called variety meats—the seal's liver, heart, and lungs. They also made a sausage of seal blood. The blubber was also eaten, along with the meat. Generally, seal meat was boiled fresh or the whole seal was put into a cache, where it fermented. This process is said to make the meat taste both strong and sweet, and the fermented livers have a flavor not unlike preserved cranberries.

In modern times, the fur of baby seals provided income for some Eskimo, but this practice has been highly criticized by some conservation groups. Widespread seal hunting, usually for oil, has in the past caused some trouble among nations regarding both the arctic and antarctic waters. However, the seal has never been used to a great extent for food in the antarctic simply because the region lacks the equivalent of the Eskimo. Seals are quite plentiful around some of the islands near Antarctica, such as South Georgia, and they are eaten in some places. But they are not as widely used for food as the arctic seals. One exception is a small tribe of Indians, the Yahgans, who inhabit the lower islands and maze of chilly waterways off the tip of Argentina. Working from canoes, they make their living from the edges of sea; the seal is an essential animal in their way of life.

These Indians are the southernmost culture of the world.

sea robins Several species of sea robins (family Triglidae, genus *Priunotus*) swim the waters of the Atlantic and the Mediterranean. All of these have large pectoral fins that resemble wings. In spite of the popular name sea robin, they aren't true flying fish. They are bottom feeders. The wings are used for walking along the bottom while looking for food. The fish will, however, come to the top, then glide back down, perhaps just for the fun of it.

Unfortunately, the sea robin isn't widely eaten in America. It is a small fish (about 12 inches long) with a large, ugly head that is covered with bony plates and spines. Possbily because it resembles the puffer, which has some highly poisonous innards, the sea robin has the unfortunate reputation—in parts of America— of being dangerous to handle and eat. The fish is easy to fillet and its delicate white meat is quite tasty, provided that the oily skin is removed.

The fish has a better reputation as table fare around the Mediterranean Sea. Several species (called gurnets or grondin) are highly esteemed, and they are often used as the main ingredient in the classic French stew called bouillabaisse. One European species, called the shining gurnet, has wings of a deep blue or purple color, and is said to resemble a butterfly. When filleted, one species of sea robin looks and tastes exactly like another.

In any case, the sea robin is definitely a neglected fish in America, although it is taken in good numbers by Atlantic sport fishermen while bottom fishing, and in large numbers by commercial fishermen, especially shrimp trawlers. Although it can sometimes be found in American fish markets, the sea robin can truly be called the "unknown delicacy."

sea snakes This group of snakes, all of which have a flattened tail, are actually more numerous than land snakes. All of the 55 species are highly poisonous—even more lethal than the cobra—but they are usually not aggressive, and their fangs are small. Many fishermen remove them from nets by hand; but handling sea snakes is certainly not recommended.

Often caught in nets, sea snakes are eaten in large numbers in Japan, China, and Polynesia. In the Philippines, they are cultivated in saltwater ponds for their meat and skin. They are also cultivated in Japan. Some seagoing fishermen keep sea snakes alive in barrels of water so that they will have a fresh supply when needed. All of the 55 species (family Hydrophidae) are edible, but most are too small to be of commercial value. Most sea snakes are 3 or 4 feet long, and some individuals attain lengths of up to 8 feet.

The Atlantic has no sea snakes, unless a few have come through the Panama Canal or possibly the Suez. They are common in parts of the Pacific, from Japan to the Gulf of California and Ecuador. In the Indian Ocean, they range all the way to the Persian Gulf. Usually, they are found in rather shallow water, often near island beaches. But some species are found in open sea, where they sometimes form schools several miles long.

There is little danger of getting poisoned by eating sea snakes, but nonetheless the *U.S. Armed Forces Survival Manual* recommends that the head be removed before eating. That makes sense, since that is where the poison is located. In spite of evidence to the contrary, one survival manual says that all snakes are edible *except* sea snakes.

(See also EELS; SNAKES.)

sea turtles All seven species of seagoing turtles grow to a large size and have been used for meat. Some peoples, such as the Andamanese islanders, hunted them with harpoons, and others caught them with nets or with hook and line. In recent times, the sea turtles have been caught in nets by fishermen trawling for shrimp or commercial fish. Although commercial fishing has taken its toll on sea turtles, the biggest threat comes from the fact that they usually lay their eggs on the shore, typically a sandy beach. Both the turtles and the eggs are quite vulnerable at this time, and loss of beach suitable for egg laying has also taken its toll. The eggs are often eaten by the local people, and have been marketed. These days, many of the egging beaches are fully protected by law.

Turtle shells have also been used by man, and the turtle industry as a whole was important at one time in some areas. In the Cayman Islands turtles are raised even today, and tourists feast on turtle steaks and soup. The Arawak Indians of the Caribbean region, a peaceful folk, were fond of turtle meat and, believe it or not, caught them with the aid of remoras. The Caribs, on the other hand, didn't eat turtle meat because they felt that it would make them stupid. (They had no such qualms, however, about eating their neighbors, the Arawaks.)

The Andamanese islanders, who didn't traffic much with the outside world, depended so heavily on the sea turtle for food and oil that the reptiles became a part of their culture. Although the Andamanese live on islands that are suitable as nesting sites, they did not take the turtles when they came ashore. Instead, they hunted them in the sea with harpoons. Their canoes had long elevated platforms in the bow, where the hunter stood ready. But he didn't merely throw the harpoon at the turtle. Instead, he held it until the bow platform was directly over (or very near) the turtle. At the right moment, he jumped off the platform atop the turtle, using his body weight to drive the head of the harpoon through the turtle's shell. The harpoon head was detachable and to it was tied a long length of rope. The harpooner returned into the canoe, and with the paddler followed the turtle until the end. Clearly, the harpooner had to have skill as well as courage.

The Andamanese caught the hawksbill turtle (*Eretmochelys imbricata*), which can weigh up to 400 pounds, and a smaller species. The world's largest turtle in the sea is the leatherback, *Dermochelys coriacea*, which weighs up to 1,500 pounds. Sea turtles have been important sources of food, partly because of their size, but other families of turtles have also been eaten.

(See also TORTOISES; TURTLES.)

sea urchins Several species of these strange creatures are eaten around the world. Known as *erizos* or *herizos* in South America, they are a popular delicacy in Chile, where they grow large along the coastline; these are eaten raw, along with chopped onion, lime juice, salt, and pepper. Erizos are also cooked in omelets as well as in other dishes. In France, where they are called *oursins*, or sea egg, they are cooked in several ways, including merely boiling them

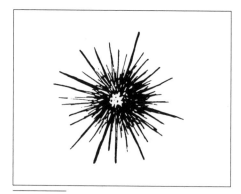

Sea urchin

lightly in seawater. They are also called sea eggs in the West Indies, where they are popular among the natives. Japan uses lots of sea urchins, or *uni*, which are eaten fresh (as in sushi) or used in making a fermented paste.

Sea urchins must be handled with caution. They look like pin cushions in the sea, and the spines can inflict a painful sting. (The amateur forager should start out with heavy gloves.) The animals have a hard shell under the spines, but their bottom sides contain a skin covering that can be opened easily. The edible eggs, usually orange or yellow, can be seen along the inside top part of the shell and are scooped out with a spoon or a cracker. Often, sea urchins can be gathered from shallow-tide pools, and they are also caught commercially by scuba divers.

About 700 species of sea urchins exist, and most mature specimens—both male and female—bear edible eggs during the spawning season, which varies with water temperature. Some species grow up to 10 inches in diameter, and others are too small to provide much food. One species, *Echinus esculentus*, is eaten in England and Europe, and *Paracentrotus lividus* is eaten in Ireland as well as in the Mediterranean.

sharks Over 300 species of sharks roam the seas and oceans of the world, swimming the shallow waters and the deep. These fish are usually associated with images of large mouths and sharp meat-shredding teeth, as depicted in *Jaws* and other movies. Not all of them are so viscious, however, and some of the bottom feeders even have flat teeth for cracking mollusks.

A number of sharks are edible and some are highly prized for food. The mako, for example, which grows up to 20 feet long, has very good flesh that has been compared to that of the highly prized swordfish. As it happens, the mako often feeds on the swordfish, which it runs down and catches in open water. The mako—a warm-blooded animal—is believed to be the fastest fish that swims, able to go up to 60 miles per hour. It is a dangerous shark, and is known to attack small boats. Mako sharks can leap as high as 30 feet out of the water, and sometimes fall on boats. There are several species of mako. Most are deep-sea fish, and the one most likely to be encountered by fishermen is the shortfin mako (*Isurus oxyrinchus*), which sometimes comes close to shore.

The porbeagle shark (*Lamna nasus*) that lives in the cold waters of the North Atlantic is also eaten and, like the mako, the texture of its flesh is often compared to that of the swordfish. The hammerhead (*Sphyrna mokarran* and other species) is also eaten, but its flesh requires proper care. The fins of hammerheads are used in soup, especially by the Japanese. The blue shark is edible, but usually not highly regarded. The thresher shark, tiger sharks, and great white shark are also eaten, along with the blackfin and many others.

In recent history, sharks have been eaten much more widely in Europe, South America, Africa, and Asia than in the United States and Canada. Both Japan and Mexico eat lots of shark. In parts of Central and South America, shark has in the past been sold in salted and dried form. Curiously, the Greenland shark is said to be toxic when eaten fresh, but it is perfectly acceptable in dried form or when decayed in the Eskimo's *tipnuk*. According to *The Encyclopaedia Britannica*, the school shark (*Galeorhinus australia*) was at one time a valuable source of vitamin A (the oil of some shark livers is especially high in vitamins), and has also been the source of "flake," used in England's "fish and chips." Other sharks have also been used in this British specialty.

The *Britannica* also points out that shark meat is usually sold under some other name, and this does seem to be the case. The terms greyfish and whitefish have been used for shark meat, and the large and numerous blue sharks do indeed have snow-white flesh. The mako shark has even been marketed as "swordfish" steaks in the United States. The reputation of the shark as table fare has changed in recent years, and the meat has become more in demand in the United States partly because of the trend toward eating more broiled and grilled fish.

Some sharks that were not eaten 50 years ago are now considered to be very good. Proper care of shark meat is essential in some cases, partly because some species contain quite a bit of uric acid, which makes their flesh smell and taste like ammonia. The ammonia flavor can sometimes be avoided if the fish is gutted as soon as it is caught and if the meat is then soaked in ice-cold salt water. As a rule, shark is better for grilling or broiling than it is for frying. Jerky can be made from the belly of shark, and many species have fins that are used for making sharkfin soup. Of course, the soupfin shark (*Galeorhinus zyopterus*) is the main species for this ancient oriental specialty. In the past, the skin of some sharks was used for shagren, a tough leather. Often, shark meat has been used for making fertilizer and animal feed.

(See also FISH.)

short-tailed shearwaters   (*Puffinus tenuirotris*) These members of the puffin family represent just one of several species that fly around the oceans of the world. Rather large birds, they measure from 15 to 25 inches long, with wingspans of up to 30 inches. They get their name from their habit of skimming over the sea just above the water. All of the shearwaters are edible. When young, they become quite fat and they are even today taken in large numbers for meat and oil. Short-tailed shearwaters are harvested commercially in Australia and Tasmania, or on the islands in Bass Strait. They are known as mutton-birds because they were so widely used as food by the early European settlers.

The mother birds lay a single egg in November, using a hole or burrow in the ground. Both parents feed the single chick. Within 6 weeks of hatching, the young squab actually gets so fat that it weights as much as or more than the parents. After the parents feed the squab for 14 weeks, they leave. The squab stays in the nest and lives off of its own fat until its feathers grow and it is ready for flight. When the squab leave the nest, they start a long figure-eight migration, lasting

three or four years. After leaving Bass Strait, they fly along the coast of New Zealand, then north to Japan, arcing east across the Aleutian Islands, circling south to Baja California and then heading southwest back to Bass Strait, where they alight on the same island on which they were born. Here they mate and start the cycle again.

Hundreds of thousands of the squab were killed annually by hunters. In the old days, they were salted and sold. These days they are canned and sold under the name of Tasmanian Squab. Because the birds lay only one egg, continuing to take such large numbers would have severely reduced the populations. The Australian government, guided by scientific study, put strict regulations on the harvest. Even today the flocks of returning shearwaters are said to blacken the water of Bass Strait every October. Apparently the population has stabilized, although many are still taken for meat and oil each year.

The world's other shearwaters are also eaten here and there along their flight patterns, but the birds don't follow ships and aren't readily available to people. Besides, it's the oily young birds that are highly prized as table fare. These are sometimes taken at nesting areas, such as the islands of the Tristan da Cunha group. Others nest off the coast of Europe and on islands as far north as Iceland.

(See also BIRDS.)

shrimps and prawns   Zoologically speaking, the terms shrimps and prawns are not exactly defined and sometimes it is hard to tell which is which. In general, a prawn is considered to be a little thinner than a shrimp, and is longer. But the two terms are used interchangeably. By whatever name, these crustaceans are quite important commercially.

The largest of the world's several species of shrimps includes the saltwater *Penaeus monodon*, which lives in the Pacific and attains a length exceeding 12 inches, and the freshwater *Macrobrachium* species, which attains a similar length and which lives in some streams of Texas and elsewhere. Shrimps are caught on a small scale by fishermen with cast nets, and on a large scale by seagoing boats that pull large trawls. Although the Gulf of Mexico and the southeastern coast of the United States are important in the shrimping industry, other areas—including the cold waters of Alaska—also produce shrimps in large numbers. Shrimps are also raised commercially in Japan and other parts of the Orient.

One of the smallest shrimps lives in small ponds in, of all places, the Sahara Desert. This is the tiny *Artemia salina*, which grows only to about ⅛ inch long and can be found in the oceans as well as in the desert ponds. In parts of the Sahara, these tiny shrimps are called *doods*, the Arabic word for worms. The local people gather these and pound them into a paste, which is eaten on bread. Other very small shrimps are also eaten, often without peeling off the shell.

Shrimps are marketed fresh, frozen, canned, and dried. Shrimp bran, made from shrimp heads and peel, is used to feed livestock. The Japanese eat raw shrimps, and the Brazilians are fond of dried shrimp appetizers. Dried shrimps and shrimp pastes are also used in oriental cookery. Most people cook only the tail of shrimp, but in Louisiana the whole shrimp is boiled. The tail is peeled and eaten, and the liver and other parts are sucked out of the head part.

silkworms A number of worms and insects produce silk in one form or another. Of course, the silkworm spins its cocoon with an extremely thin filament, which can be up to 1,000 feet long. In silk manufacture, these filaments are unwound from the cocoons and spun into silk threads. Such use of silkworm cocoons is believed to have been started in China about 2000 B.C. by Empress Siling. In China, silk is called *si*. The silk industry grew in China and the wondrous new material's fame spread around the world. Although the Chinese tried hard to keep the secret (disclosure of methods of production was punishable with death by torture), ancient Sanskrit writings also refer to commercial silk production in India. Aristotle spoke of the silkworm's cocoon and the spinning of silk. It is said that a Byzantine emperor persuaded two Persian monks to smuggle silkworms out of China in hollow bamboo canes. From Byzantium, silkworm culture spread throughout Europe. Eventually, Japan became the most important producer of silk. These days, however, the mechanized manufacture of nylon and other materials has reduced the demand for silk.

Most commercial silk is made from the worm of a moth (*Bombyx mori*) that feeds on the mulberry tree. The worm has become completely domesticated, and probably no longer exists in the wild. Other moths also produce cocoons of silk, some of which feed on oak and other trees.

Since insects are eaten by several peoples, it is not surprising that the silkworm should also be consumed. According to Marco Polo, silkworms were sold in the markets of Hangzhoo, the capital of China during the Middle Ages. Actually,

---

Missouri Parboiled Skunk

Skin, clean, and remove the scent glands. Put in a strong solution of salt water and parboil for about 15 minutes. Drain off this water, add fresh water, season, and steam slowly for about 1 hour or until tender.

—*Cy Littlebee's Guide to Cooking Fish & Game*

---

the part used for food was the pupa that was left in the cocoon after the silk filaments had been unwound.

(See also GRUBS AND WORMS; INSECTS.)

skates See RAYS AND SKATES.

skunks These strictly American members of the weasel family make good pets when properly deodorized. The more common species is the common striped skunk, *Mephitis mephitis*, which lives from Canada to Mexico. It is trapped and raised for fur, and, believe it or not, its scent, which can linger in clothing for weeks, is of some value in the perfume industry. The skunk is not exactly daily table fare in America, but it is sometimes eaten and highly praised. A report by Waverley Root in *Food* quoted a descendant of the Mashpee Indians of Massachusetts as saying that cranberries were made to go with roast skunk instead of with turkey. Also according to the same book, William Byrd in the 18th century said that the polecat was surprisingly sweet. Byrd was no doubt talking about the skunk.

A polecat recipe was included in *Cy Littlebee's Guide to Cooking Fish & Game*, published by the Missouri Department of Conservation. According to an account sent in by a fellow from St. Louis, the skunk has the sweetest taste of all the wild game. After setting forth the recipe, Cy said to the reader, "And if you do try it, let me know how it turns out."

sloths Arboreal and nocturnal, these strange creatures spend a good deal of the time upside down. They almost never leave the trees, but they don't run along the limbs like squirrels and raccoon. Instead, they cling to the limbs, underslung, as they go. This explains the curved shape of their claws and the absence of pads or soles on their feet. Sloths can get about on the land if necessary, but they are quite awkward at it.

There are two families of sloths: Megalonychidae, or two-toed sloths, and Bradypodidae, or three-toed sloths. Actually, both kinds that have three "toes" on the hind legs. It's the "fingers" that differ in number. Both kinds are vegetarians that inhabit the tropical forests of Central America and the northern half of South America. The two-toed sloth is the larger of the two, weighing up to about 17 pounds.

An early Spanish writer, Oviedo y Valdés, said that he had never seen an uglier or a more useless animal. Ugly or not, the South American people hunt the sloth for meat and eat large numbers of them. The prehistoric giant sloth was also eaten, and may have been hunted to extinction.

snails Somewhere between 60,000 and 75,000 species of snails and slugs exist in the world, on land and in the sea. Some of these have been eaten since ancient times; during the Stone Age snails were consumed in large numbers by the primitive peoples in the Pyrenees, as evidenced by huge deposits of snail shells. The most famous snail, of course, is the escargot, which was relished by the ancient Romans and currently by the modern French and gourmets everywhere. These are called Roman land snails, both *Helix pomatia* and *Helix aspersa*, which were raised in special gardens called cochlearia and fattened on milk. The *Helix aspersa* is the common garden snail, which grows in France, England, North America, and other places.

The Chinese are also especially fond of land snails, especially in the Canton area, where they are sometimes roasted. A large land snail is important food to some of the local people in parts of Africa. In addition to the land snails, slugs are also eaten here and there, and only recently have been proposed as a major source of food. (Slugs are really snails without shells, and some grow up to 9 inches long.) Also, sea snails such as the abalone, the conch, and the periwinkle are also widely eaten in various parts of the world.

Snails are also used for making dye. The Murex snail was once an important item of commerce for the peoples of the Mediterranean and was used in a famous dye made by the Phoenicians. The shells have been widely used as vessels and as pearl-like material for inlays and jewelery. The shells have also been used as money and as horns.

(See also ABALONE; BIVALVED MOLLUSKS; CONCHS; PERIWINKLES.)

snakes Some snakes may be more highly prized than others as table fare, and surely there are regional favorites, but all snakes are edible. In the American South and West, rattlesnakes are frequently eaten and, in the recent past, rattlesnake meat has been canned for the market. When properly prepared, it is good, but the large older snakes can be tough and are hard to eat when merely fried.

Snakes are eaten in various parts of the world, often by primitive peoples, such as the Australian aborigines, who wrapped the reptiles in clay and baked them. Today, snakes are often eaten in the Orient, especially in China, where they can be purchased in markets or in restaurants. In Canton, a traditional dish is called "the dragon and the tiger," in which the dragon is represented by the snake. The Chinese eat the python pickled in vinegar, and dried snakes are sold at markets in Hong Kong. Even the cobra and the giant boa are eaten. In France, a dish called hedge eel is really made with a common green snake.

Some writers, including William Bartram, have recorded that snakes that have bitten themselves cannot be eaten safely. This is a myth that isn't likely to die, at least not in the Western world.

There are many state programs in the United States that attempt to regulate the killing of snakes, especially those species that are considered endangered. But the battle will be hard to win, and the old saying, "The only good snake is a dead snake" will more than likely prevail in the forseeable future. Anyone who wants a mess of snake meat, however, had better first check the local laws before going hunting.

> ## Desert Survival
>
> Sand boas stay in their burrows during the heat of the day and come out in the evening to feed on small mammals or lizards. Some refugees of the Russian Revolution fleeing across Asian deserts are reported to have survived only through the sudden nightly appearance of this source of protein.
>
> —*The Encyclopedia of Reptiles and Amphibians*

Several annual rattlesnake roundups, usually fund-raising events held by some civic group, are held in parts of the American South and Southwest. Until recently most of these events sold cooked rattlesnake meat as well as belts and snakeskin products. But this practice is on the decline because of pressure from various groups that want to curtail the hunting of snakes.

spiny dogfish (*Squalus acanthias*) This relatively small shark of the North Atlantic and the North Pacific usually grows only 2 or 3 feet long. It moves about in large schools of many thousands, and sometimes clogs the nets of fishermen. Its flesh tastes good, and Europeans eat it frequently. But the shark has not gained commercial status in the United States, although it has been used in the manufacture of fertilizer. The skin of spiny dogfish has been used by turners to polish ivory. The skins of this and other sharks have been used extensively for leather.

(See also FISH; SHARKS.)

springhares (*Pedetes capensis*) The only members of the family Pedetidae, these animals look rather like large ground squirrels with rabbit ears and the legs of a miniature kangaroo. They live on the grassy plains from South Africa north to Kenya. They feed mostly on grass, although they sometimes eat dirt that is rich in minerals. The animals weigh up to 9 pounds and are valued for their meat.

Springhare

Although the springhare can be a pest to agricultural peoples, it is also an important source of food for some of the bushmen. In fact, according to *The Encyclopedia of Mammals*, the springhare is the most important wild animal in the human diet in Botswana:

No part of the dead springhare goes to waste. Over 60 percent is eaten, including the eyes, brain, and contents of the intestine. The skin is used to make bags, clothing, and mats; the long sinew from the tail is used as thread; the fecal pellets are smoked.

squid See OCTOPUSES AND SQUID.

squirrels These rodents comprise 267 known species and live on most lands in the Americas, Asia, Europe, and Africa. Some squirrels live primarily in trees, where they feed and nest. Others live in dens in the ground. All squirrels are edible, but some are mouselike in size and not of much importance. The largest member of the squirrel family is the edible Alpine marmot, which is a first cousin to the American groundhog (or woodchuck) and which may weigh up to 17 pounds.

In America, the gray squirrel (*Sciurus carolinensis*) is often hunted for sport and for meat. It has been introduced to England, where it is now widespread. The larger fox squirrel (*Sciurus niger*) is also widely hunted. The tree squirrels are active by day, which make them a popular target for rifle marksmen. They are also tasty, especially the young ones. Indeed, squirrel hunting was something of a rural tradition in America; such important figures as Thomas Jefferson were squirrel hunters. The sport is still popular, and squirrels are quite plentiful. In Illinois alone, some 1.5 to 2 million squirrels are bagged each year.

A famous American dish, Brunswick stew, was probably developed with squirrels and homegrown vegetables somewhere east of the Mississippi. Either Brunswick, Georgia or Brunswick, Virginia may be the place of origin—but nobody knows for sure.

(See also PRAIRIE DOG; RODENTS.)

stone crabs (*Menippe mercenaria*) The stone crab has a hard shell, and is rather thick and stocky as compared to other crabs. Although it grows to only about 5 inches in diameter, it has large claws; and these are the only parts that are legally kept and marketed. These claws or pincers are very thick, hard and difficult to crack open without a hammer or pliers. They are very colorful—coral and black—and the whole claw is usually cooked, either by steaming or boiling for a few minutes. The meat is tasty and has a good texture; the muscles, designed for working the pincers, are made up of short fibers, and are not long and stringy like those in some of the long legs of other species of crabs. In fact, stone crab is probably the most highly prized of all the crabs.

In the United States, the stone crab lives from North Carolina to Texas, and is more important commercially in south Florida than other locations. At one time, the numbers of stone crabs were drastically reduced by commercial harvest, as well as by foragers after the crab for their own consumption. Fortunately, it was discovered that adult stone crabs could survive with only one claw; if a claw is removed, the crab will grow another. Thus, sport and commercial fishermen who catch a two-clawed stone crab of legal size must carefully remove one claw and return the animal to the water.

(See also CRABS.)

sturgeons Most sturgeons are very large, primitive fish with bony plates and scutes. Approximately 20 species are known. Typically, they live part of their lives in the seas of the Northern Hemisphere and run up freshwater rivers to

Azerbaijani Kabob

Some of the best sturgeons come from the Caspian Sea, and they are much enjoyed by the peoples of Azerbaijan. Since the sturgeon has a firm flesh, it is suited for grilling or for cooking on skewers, as in the following:

2 pounds (907 grams) fresh sturgeon meat
1/4 cup (59 milliliters) sour cream
1/4 cup (59 milliliters) fresh lemon juice
1/4 cup (59 milliliters) melted butter
salt

Cut the sturgeon meat into 1-inch (2.5-centimeter) cubes and sprinkle it with salt. Mix the sour cream and lemon juice, then coat the fish cubes with the mixture. Save the remaining mixture and stir it into the melted butter, making a basting sauce. Thread the cubes onto skewers and grill over a hot charcoal fire, 4 inches from the heat, for about 10 minutes, turning and basting from time to time.

spawn. Sturgeon have, in the past, been found in large numbers in the rivers of southern Russia as well as in Siberia. Years ago, they were common in the rivers of France, but they are no longer taken on a regular basis. They have also been greatly reduced in most other European rivers.

The flesh of the sturgeon is highly regarded in some areas, but not in others. Some gourmets consider it to be a delicacy when it is properly smoked. For the most part, however, sturgeon roe—used for making caviar—is far more important commercially than the meat. (Strangely, in Iran the eggs are of little culinary value to some of the local peoples, although they might be exported at high prices.) The female sturgeon can produce more than a million eggs, which can amount to as much as 25 percent of her total body weight at spawning time. Since the beluga species (*Huso huso*) of the Caspian and the Black seas, the Sea of Azov, and the Eastern Mediterranean can weigh up to 2,500 pounds, one fish can in theory produce over 600 pounds of caviar.

Several sturgeons live in North America or ply American rivers during spawning time, and they were once quite plentiful. In recent years, however, they have been greatly reduced. One of these,

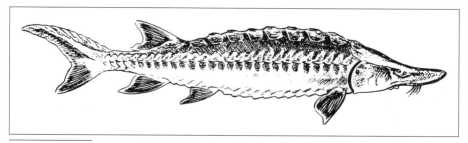

Atlantic sturgeon

the lake sturgeon (*Acipenser fulvescens*), is a freshwater species that ranges from Hudson Bay south to Northern Alabama.

(See also FISH.)

swans Comprising seven species, swans are the world's largest waterfowl. They are quite graceful in the water, where they feed on aquatic vegetation. Having short legs, swans are a little awkward on land. The mute swan (*Cygnus olor*) was domesticated in England before the 12th century and was considered to be the property of the Crown, which had a "swan master"; the swans could be raised for meat and show, but only by permission of the Crown. A native of northern Europe, the mute swan was introduced to other countries, and it is still the most popular tame swan, often seen in private ponds and recreational areas.

The large trumpeter swan, a wild North American bird, is now fairly common in some western national parks. Its range is from Alaska to Texas. Also, the wild whistling swan ranges from Alaska to Hudson Bay, south to Currituck Sound in North Carolina, where it was at one time a popular game bird during the winter season. Other species include the striking black swan of Australia and Tasmania. All of the swans have been eaten by hunting peoples, and of course they have also been raised for food. They were popular for state banquets in medieval Europe, partly, no doubt, because of their size. They were plucked, cooked, and then fitted with their original feathers to make a show at the table.

The bones of swans have been found in the refuse heaps of prehistoric man. Some of these have been traced back, by carbon dating, to the early Neolithic Age.

T

tapirs Three species of this large woodland mammal, family Tapiridae, live in Central and South America. Somewhat surprisingly because of the sheer size of the Amazon rain forest, the tapir is the largest wild animal in Brazil. Another species, *Tapirus indicus*, lives in Southeast Asia and is the largest of the tapirs, weighing up to 650 pounds. All of the species are quite similar, except for coloration. The body of the tapir resembles a pig and, indeed, the flesh is similar to pork and is highly prized. The animals seem to have no neck, and their heads are snoutlike.

Tapirs forage at night, roaming widely and feeding on grass, fruits, leaves, and other vegetation. They sometimes cause some damage to crops. Often the tapirs feed on aquatic vegetation, and they spend a good deal of leisure time in the water. They will dive under the water when threatened; the Malay tapir can even walk on the bottom like the hippopotamus. They are, however, primarily land animals, preferring thick woodlands and forests with plenty of good water holes. The mountain tapir (*Tapirus pinchaque*) lives in the Andes, up to the timberline, in Peru, Equador, and Colombia.

The tapir is hunted for food and sport. It is also prey to the jaguar, the leopard, and the tiger. The tapir's thick hide, without much hair, is used to make leather that is especially tough. But the biggest threat to the tapir is the destruction of its habitat by logging and agricultural expansion.

tarpons The Atlantic tarpon (*Megalops atlantica*) is a great warm-water game fish, weighing up to 350 pounds, and is known for its ability to leap and walk across the surface on its tail when it is hooked. The Pacific tarpon is quite similar, only smaller, and the tropical 10-pounder is still smaller. The tarpon has large, thick scales, which have been used in making jewelry. The fish is commonly eaten along the coasts of Central America, but is not usually eaten along the Gulf coasts of the United States. The roe is considered to be a delicacy in some quarters.

Also called sabalo reel, cuffum, silverfish, and silverking, the tarpon is a bony fish. Its flesh gets mixed reviews. *McClane's New Standard Fishing Encyclopedia* notes that it has little food value, except that the roe is exceptionally good. One New Orleans cook, Frank Davis, author of *The Frank Davis Seafood Notebook*, makes no bones about his feelings concerning tarpon. He believes that the fish has very tough, coarse, pungent meat, with a high oil content. He reports that some experts claim the roe can be a delicacy when panfried, but he doesn't eat it at all. By contrast, *The Encyclopedia of Aquatic Life* says that tarpon is a highly prized food fish.

The Atlantic tarpon ranges up and down the Atlantic, going as far north as Cape Hatteras and as far south as Brazil. It is much more common, however, in Florida and Gulf of Mexico waters and in Central America. It also lives in the Eastern Atlantic off the coast of Africa. The tarpon frequents open water as well as bays and brackish waters, and it even goes up freshwater rivers for considerable distances.

(See also FISH.)

Termites à la Bantu

The members of the Bantu tribe of South Africa eagerly collect swarming termites by placing pails of water under the nest opening, into which many of the insects fall and drown as they emerge from the nest. These are used for food and here is a rare recipe for "Termites à la Bantu":

  1 pint (half a liter) termites
  1 teaspoon (5 milliliters) vegetable oil (ground of palm nut)
  salt, if available

Remove termite wings. Spread on flat stone in sun to dry. Smear pan or stone with oil and spread dried termites upon it. Toast over hot coals until almost crisp. Sprinkle with salt. Eat like popcorn immediately or store for future use—thay can be stored for months!

—*Insect Fact and Folklore*

---

tenrecs This group of insectivores live only on the island of Madagascar, where they are eaten for food. The largest of the group, the tailless common tenrec (*Tenrec ecaudatus*), weighs 2 or 3 pounds and has been an important source of food and cooking oil since ancient times. They make excellent eating. Some of the species, including the common tenrec, have spines and can ball up like hedgehogs, with which they are sometimes confused.

(See also INSECTIVORES.)

termites These small insects of the order Isoptera are eaten in some lands, where they are often caught in swarms as they leave their nest or mound. Some species live in rotting wood and others live in the ground. Most of the world's termites are tropical and feed on dead vegetation (usually leaves, grass, or wood), but a few feed on the organic humus of the soil and fungi. Some termites build huge mounds, often as high as 18 feet, above the ground. A large mound can house over a million termites. Some of the mounds have a hard outer shell and are difficult to crack open, even with the aid of an ax. In South America, these mounds have been hollowed out and used as ovens. Usually this "clay" is termite fecus which, nonetheless, is sometimes chewed like tobacco.

In any case, termites are rich in protein, and in the past have been a significant part of the diet of some peoples. Reportedly, they taste much like pineapple. The queen termite makes the best eating and is sometimes sought out. She becomes so big with eggs that she can't walk, and may become 50 times the size of an ordinary termite. In the East Indies, it is believed that if old men eat the queen termite their backs will be strengthened. But, as a rule, it is the enormous swarm of termites that is of interest. In Africa, the Akoa Pygmies make good use of the insects, as demonstrated by the following report from Carleton S. Coon's *The Hunting Peoples*:

As the termites tend to build their hills more or less in the same place each year, the Pygmies know where they are without search, and each man in the camp marks out his own hill. From time to time as the season approaches he goes to it and cuts into its side to see how high up the termites have risen, and thus can estimate how soon they will swarm out to fly away. The whole band may desert its huts for several days for the swarming. Because the insects begin to fly once they have come out, the people must be there beforehand. Near its hill each family sets up a windscreen, and they build a roof of leaves over the hill, and dig a deep trench around the base.

They build a fire in front of the windscreen, facing the hill, and dig a hole in front of the fire. At dusk the ants begin to fly. Striking the roof, they tumble into the trench, crawl toward the fire, and fall into the hole, out of which women scoop them up in baskets. Usually they roast the termites, although sometimes they eat them alive, pound them to a paste, or boil them. The Akoa Pygmies boil as many as they can, and scoop off the oil that rises to the suface of the water. They use this oil both in cooking and as a pomade, mixed with red wood-powder and applied to their bodies.

(See also INSECTS.)

thrushes Which bird is or isn't called a thrush varies greatly from one location to another. Generally, however, thrushes are fond of hopping about on the ground, often around hedges and low bushes; most of them are also strong fliers, and some even migrate over long distances. Since all of the many thrush species are edible, perhaps exact classification is best left to scientists and bird-watchers. To be sure, some of the best birds for the table are thrushes, such as the European song thrush (*Turdus ericetorum*), a brownish bird with a speckled breast. Song thrushes and others are highly regarded in France, where several recipes are used to prepare them, perhaps the most extravagant of which involves stuffing them with truffles. Another species, the North American robin (*Turdus migratorius*), is also considered to be an excellent bird for the table.

The ancient Romans kept thrushes in cages and fattened them for their private table—and for the market—on millet, crushed figs, and flour. Today, various thrushes are commonly eaten in some countries, but they are protected by law in others.

(See also BIRDS.)

tinamous (family Tinamidae, order Tinamiformes) About 40 species of these partridgelike birds live in the forests and grasslands of Central and South America. Growing from 8 to 21 inches long, they make excellent eating. They nest on the ground, and their glossy eggs—which may be either green, blue, yellow, or purplish brown—are also edible.

toadfish There are several species of toadfish that live in the warm waters of the world. Most of these are small, and some are kept in home aquariums. A few species have hollow spines on their fins or gills that can inject a poison, and should be handled with caution. One species, the oyster toadfish (*Opsanus tau*), more or less sits on the bottom and really does look like a toad. It can be found in shallow salt waters from Maine to Florida, and several of its cousins live farther south and in the Bahamas. The oyster toadfish grows to a weight of about a pound, and is sometimes taken by anglers fishing on the bottom for flounder and other species. The oyster toadfish is edible, but its spines should be avoided.

tortoises These land turtles are mostly tropical and subtropical, but some do survive in colder areas. All together, there are 41 species of true tortoises, most of which are vegetarians. All of the tortoises are edible and some are highly prized as food, but some of them are too small to be of much value except on a survival basis. By contrast, the Aldabra giant tortoise or elephant tortoise (*Geochelone gigantea*) weighs up to 560 pounds. It is found only on a group of small islands in the Seychelles, where it is now protected.

Another very large tortoise (up to 500 pounds) lives on the Galapagos Islands, about 500 miles off the coast of Ecuador. During the heyday of sailing ships, thousands of these large animals were eaten by whalers, pirates, and other seafarers. The large tortoises make excellent eating, and could be taken aboard the ships to provide fresh meat at sea. The tortoises of the Galapagos are now protected by law, along with other endangered species on the islands. In fact, the government of Ecuador established wildlife sanctuaries on the islands in 1935 and again in 1959. The Galapagos Islands are widely scattered and quite isolated, which allowed different strains of these turtles to develop on different islands. These variations greatly influenced Charles Darwin's theories of evolution and similar scientific thinking.

Other tortoises are also eaten. In the southeastern United States, for example, the gopher tortoise (*Gopherus polyphemus*) was at one time used for food on a regular basis. During the lean years of the Great Depression in the 1930s, the

name for the animal was shortened to gopher and they were sometimes called "Hoover hens." During this time, the art of gopher pulling was developed, based on this tortoise's habit of digging holes deep into the ground for use as dens. Typically, a gopher hole is made in dry, sandy soil and slopes down at about 30 degrees. It may be 30 or 40 feet in length. The mouth of a hole made by a fully grown gopher tortoise is about 12 inches wide and 8 inches high. Just outside the mouth a pile of dirt forms from the matter slung out of the hole by the animal's hind legs. The gopher stays in its hole at night and, during the early morning, ventures out only after the sun dries the dew off the grass. (The gopher hates water and will in fact catch a cold and die—a fact that ought to be remembered by anyone who captures one for a pet.) During the heat of the day, the gopher will likely be out of its hole, feeding on grass, berries, and other vegetation.

While the tortoise is out of its hole, it is very slow and can be easily caught by hunters. The old gopher pullers, however, looked for the holes and picked a time when the gopher was likely to be at home. A fresh hole, of course, would have lots of gopher tracks leading in and out. Once he located a fresh hole, the gopher puller would insert a special hook—usually tied onto a length of grapevine—into the hole and jiggle it around until it would catch under or over the gopher's shell. The hook didn't stick into the gopher; rather, it took a hold between the upper and lower platens. Once hooked, the gopher was pulled out either gently or forcefully. Some old gophers could push against the sides of the hole with their strong legs and resist the effort. It has been said that some simply couldn't be pulled—even with the aid of a Model A Ford.

Gophers were also taken by snares and other kinds of traps. In the sand hills of Central Florida, gophers were quite plentiful and hard-working pullers could earn pretty good money by catching them for the market.

Today, the gopher tortoise is protected in most of its range. Its biggest threat is from loss of habitat and from automobiles. It became apparent to wildlife experts, as the gopher tortoise's numbers decreased, that the holes are very important for other forms of wildlife, such as foxes, rabbits, burrowing owls, gopher frogs, and rattlesnakes.
(See also TURTLES.)

totuavas (*Cynoscion macdonaldi*) These fish grow only in the middle and upper end of the Gulf of California and are called *totoaba* in Mexico. The totuava is considered to be an excellent food fish, and it was at one time fished almost to extinction, partly because it was in great demand in the Orient, where its air bladder (sometimes called a sound or swim bladder) was a main ingredient in a soup. (Sounds of codfish are also a favorite in Newfoundland. Sounds from sturgeon were once used as a source of the isinglass used in the clarification of wine and other beverages.) In any case, the totuava is a large sport and commercial fish, growing up to 200 pounds or better.

The totuava has a habit of spawning in the north end of the Gulf of California at the mouth of the Colorado River in spring and early summer, when the moon is right to produce high tides. The fish spawn in shallow water, where they were at one time speared wantonly. In recent years, the commercial fishing season has been closed on the totuava during spawning season and the upper Gulf is protected. As a result, the species is now gaining in numbers.

From a culinary viewpoint, totuava are good eating. In fact, these fish are often eaten raw in the popular Mexican and South American dish *seviche*, the Latin America answer to Japanese sushi. The totuava is a cousin to the weakfish.
(See also FISH.)

tripletails (*Lobotes surinamensis*) Often called buoy fish or buoy bass, tripletails like to hang out around buoys, floating debris, or other objects in the water. For that reason, anglers along the Gulf of Mexico sometimes build floating rafts of brush or other material to attract the tripletail, and the Japanese go a step further and build large bamboo structures called *tsuke* rafts to attract this and other species of fish. The tripletail is so named because its top and bottom fins are long on the ends, making the fish look as though it has three separate tails.

The tripletail is widely distributed in the Atlantic, Indian, and Pacific oceans, usually in tropical and temperate waters. It reaches up to 50 pounds and is considered to be an excellent game fish. It is also highly regarded as table fare; being medium oily, it can be cooked in a variety of ways.
(See also FISH.)

tropic birds (family Phaethonidae, genus *Phaethon*) Three species of these beautiful seabirds roam the warmer parts of the Atlantic, Pacific, and Indian oceans. All of the tropic birds have slender tail feathers that are about a foot long. The birds are from 16 to 20 inches long, not counting the tail. Because the birds stay out to sea, are not gregarious and do not make a habit of following ships, they are not often taken for food except at their island nesting grounds.

When they do nest, however, the tropic birds have been hunted excessively for their beautiful tail feathers. The birds lay a single egg, usually on a ledge in a seaside cliff and sometimes under trees. The egg is not protected or even hidden, so the bird sits on it; the tropic bird does not easily run off, even when approached by man. The birds squawk and peck at intruders, but they won't budge until they are lifted off the egg. Because of this habit, the natives of some South Sea islands learned to snatch the coveted tail plumes out of nesting birds without killing them. In the past, the eggs have been used for food, but the bird is now protected in some of the nesting areas, such as Bermuda and some islands in the West Indies.

(See also BIRDS.)

trumpeters These unusual edible birds from the jungles of northeastern South America got their name because of the sound they make. There are several species, differing mainly in coloration. The most well known species, *Psophiidae crepitans*, lives in Guiana and is the size of a barnyard fowl, but with a longer neck and longer legs. In temperament and habit, as well as in its use as table fare, the trumpeter is similar to the Guinea fowl of West Africa.

(See also BIRDS.)

turtles There are 254 species of turtles in the world, including the dry land tortoise; all are edible and have probably been eaten at one time or another. Some are too small to be of much value as food,

except on a survival basis, but others are quite large and one catch can provide 100 pounds of meat or more. The freshwater turtles include what are sometimes called terrapins, a confusing term that is applied to several freshwater species. (The term tortoise is also applied to all kinds of turtles, but in this work it is considered to be a dry land reptile.) Some turtles are also called cooters.

The meat of turtles has been described as having seven different flavors, including fish, ham, and chicken. The flavor of most turtles is indeed good, but the texture of the meat is often what sets it apart from less desirable meat. The turtle has a unique bone structure, and in order to make everything work the muscles are short as compared, for example, to the long muscles in a bullfrogs's hind legs. In any case, turtles have always been eaten by man, partly because their meat is good, partly because they are easy to catch, and partly because of their nesting behavior. Sea turtles are especially vulnerable when they come ashore to lay their eggs along the edges of the beaches. The same can be said of river turtles and lake turtles, most of which lay their eggs close to the water.

The large snappers of freshwater rivers and lakes of the United States are often eaten, and are trapped on a more or less commercial basis. The many large impoundments built around the country, as well as untold numbers of farm ponds, have increased the snapper's habitat, although in some areas constant trapping or fishing may have reduced the numbers of very large, old turtles.

The two species of freshwater snappers include the alligator snapping turtle (*Macroclemys temminckii*), which grows up to 200 pounds. It has a ridge atop its back, a feature that makes it look somewhat like an alligator. This species has a large head and wide mouth. Inside the mouth grows a small wormlike organ, which the turtle uses as a fish lure.

Softshelled turtles include 22 species that inhabit the temperate and tropical fresh waters of North America, Africa, Asia, and the Indo-Australian Archipelago. They all live in freshwater lakes and rivers, but some of the larger softshells have been known to venture far out to sea. In addition to having leathery shells, which are usually hard in the middle and flexible around the edges, these turtles are flat like pancakes and have long necks. The meat of most species is very good and sometimes the softer parts of the shell are eaten, after long stewing. The softshell sometimes uses its long neck like a periscope, so that it can look about without breaking the surface of the water. It also buries its flat shell in the bottom of the lake or stream, so that only its neck and head stick out. Typically, the softshell is quick in the water, and with its long neck it can strike like a snake. In Florida, the softshell is often caught by fishermen after catfish with trotlines, and it is sometimes available in fish markets and even supermarkets.

Some writers on culinary topics talk about eating turtle flippers (as on the huge leatherback sea turtle), just as others consider frogs' legs to be the only edible parts or at least the only parts worth fooling with. The meat enclosed by the turtle shell is always good, although it is sometimes hard to get at. The so-called variety meats are also good. The soft-shelled turtle, for example, has a delicious flat liver, mild in flavor. Undeveloped turtle eggs are considered by some people to be culinary prizes and are often cooked whole in turtle stew or soup. In South America, one tribe makes a stew of turtle innards, cooking it in the shell.

(See also SEA TURTLES; TORTOISES.)

V

viscachas See CHINCHILLAS AND VISCA-
CHAS.

W

wallabies See KANGAROOS AND WALLABIES.

wapiti (*Cervus canadensis*) The name *wapiti* is derived from the Algonquian and refers to a large deer, which is called elk in America. The wapiti is one of the world's largest deer, second only to the moose. In any case, the wapiti makes very good venison and is highly regarded both for sport and for meat. It stands over 5 feet high at the shoulders, weighs up to 1,000 pounds, and has a large, impressive set of antlers.

In addition to its range in northwestern North America, it also lives in parts of Asia. The wapiti is an important source of food in parts of Siberia, where a variety called *maral* is bred on farms. (See also DEER.)

warthogs (*Phacochoerus aethiopicus*) These ugly animals are widely eaten in Africa, where they live everywhere south of the Sahara. They weigh up to almost 250 pounds, and feed mostly on grass. Contrary to popular reports, warthogs don't usually root for food or use their curved tusks for digging; they graze like cattle.

water boatmen Aptly named, these aquatic insects have bodies shaped rather like boats and their legs are used much like oars to move them through the water. Actually, however, several kinds of insects are called water boatman or back swimmers. The ones referred to here are of the Corixidae family. Sometimes occurring in very large numbers, they live on or close to the bottoms of pools of water, where they feed on the muck that contains both plant and animal matter.

The water boatman lays eggs en masse on any submerged object, and these eggs are edible. In parts of Mexico, where the insect is called *axayacat*, the eggs are sometimes marketed as caviar. To get the eggs in quantity, the Mexicans insert woven rush mats into the water at suitable spots. The mats are later removed, and the water boatmen's eggs are scraped off, sifted to clean out the silt, and put into sacks. It takes about 200 million eggs to make a pound.

(See also EGGS; INSECTS.)

whales Primitive man has hunted whales for meat since the beginning of recorded history, and Eskimo kitchen middens of 3,500 years ago contain whalebones. Whaling became a way of life for some peoples, and even in modern times a few tribes of Eskimo have depended almost entirely on whale meat and oil.

Often the whale hunt involved the whole village. Since hunting the whale was a complicated business and required a number of people—and often several boats—the meat had to be divided at the end of a successful hunt. There were several ways of division, but in every case a rigid system had to be followed, lest the whalers fight among themselves. Since some parts of the whale are better than others (just as T-bone beefsteaks are better than round steaks), the animals were cut up to exact specifications. As a rule, the solid meat that starts about a foot below the whale's naval is considered to be the best. Sometimes choice parts are eaten as soon as the hunt is concluded. Some of the Alaskan Eskimo, for example, quickly consume the meat atop the whale's head, and the whole village takes part in eating it; also, the tail and the last part of the body that joins the tail are saved for festivals.

In addition to the Eskimo, the Japanese eat lots of whale meat, and the Scandinavians have used it in the past. As a group, Europeans don't think too highly of it, although whale tongue has received high marks. According to *Larousse Gastronomique*, whale meat was consumed by the poor during the middle ages, when it was called *crapois* or Lental bacon. At that time, whale oil was important for use in lamps, and large numbers of right whales were taken by the Basque hunters, who had the European market more or less cornered for several centuries. By modern standards, however, the meat is not highly regarded. It is reddish in color and is said to resemble tough beef that has been boiled in water in which mackerel had been washed.

Nevertheless, the meat has been crucial in the diet of some coastal peoples, who depended on the whale for oil, bones, and meat. Other people hunted the whale but also had other food available; in these cases the dangerous hunt was more of a sport and a test of mettle. Many other peoples ate whales whenever they found them beached along their shores; some tribes might well have made temporary camp near the whale until it was consumed.

Among the primitive peoples, hunting whales was mostly a matter of killing enough (and no more) than the immediate tribe could use, although there might have been a little trading here and there. The meat and the oil were of utmost importance, but the whalebone and teeth were also used. Evidence indicates that even houses were sometimes made of whalebone in parts of Greenland. In Scandinavia whale vertebrae were used as stools.

The Eskimo and other primitive peoples hunted the whale using harpoons from small boats. Lines were attached to the harpoons, and the lines were connected to floats made of inflated sealskins. The floats allowed the hunters to follow the wounded whales to the end. Sometimes the harpoons of the early Aleutians were dipped into a poison to speed things up, and some other peoples dipped their harpoons into the blood of dead whales, which was believed to cause blood poisoning in the live whales.

The hunters had a big problem if a dead whale sank; for this reason those whales that stayed afloat after death were called right whales, simply because they were the right ones to hunt. Although other species were surely taken and were eaten at every opportunity, it was the right whales that were most important to the early hunters. These included two species: the bowhead whale (*Balaena mysticetus*) and the right whale (*Balaena glacialis*), both of which are now endangered. These are large whales, weighing from 50 to 80 tons. Often, they swam close to shore and, it should be noted, would typically feed by skimming the surface with their mouths open, making them easy to locate and harpoon.

The Norsemen may have been whalers to some degree, but the first really big commercial whaling operations were developed by the Basque peoples of Spain and France. The right whales frequented the Bay of Biscay, and that is where the whaling industry began in the 12th century. Before long, the whale oil was sold all over Europe and the whale population in the bay began to thin. The Basque, of course, made larger boats and hunted the whales at sea, going all the way to Iceland and maybe Greenland and Newfoundland. According to *The Whale* by Jacques-Yves Cousteau and Philippe Diolé, a tombstone bearing a Basque inscription has been found in Newfoundland, dating to the end of the 14th century—a century before Columbus reached the shores of North America.

Following the success of the Basques, the English, Dutch, Danes, Americans, French, the Japanese, and other peoples got into the act and whaling became big business. The meat, however, was not of much importance. These whalers were after oil (which found its way into margarine and other food products) and, sometimes, various other products. For instance, spermaceti, a wax found in the chamber of a sperm whale's head, was quite valuable at one time, and a large whale yielded a ton of it.

In modern times, the whale industry has been more or less dominated by Russia and Japan. The Japanese use meat from baleen whales for human consumption, but the Russians are interested mainly in the oil from sperm whales.

All of the 76 species of whales and dolphins are mammals, order Cetacea. (See also DOLPHINS AND PORPOISES; MANATEES AND DUGONGS; NARWHALES.)

whitebait This term refers to tiny fish that are used as food. It doesn't refer to a specific kind of fish, although it has been used in some markets to designate several species of small fish or the fry of larger fish.

The practice of cooking small fish—no longer than 2 inches—has no doubt been around for a long time, but it became something of a fad in England during the 18th century and caught on elsewhere among epicures. During the reign of King George IV, the royal household was reportedly supplied with whitebait every day of the year. This British dish was usually made of the fry of smelts, herrings, sand eels, and so on. Ideally, the dish contained several species of saltwater fish, and usually these are fish that school in large numbers in the shallow waters of the North Sea. Some are found in streams during runs. Small sardines, silversides, and many other fish have also been marketed under the name whitebait.

The tiny fish are cooked by shaking them in flour, putting them into a basket, shaking off the excess flour, and deep-frying them in hot oil for about half a minute. (The oil should be heated almost to the smoke point, and not too many fish should be in the basket at any one time.) The fish are then drained and eaten bone and all. Fish longer than 2 inches might require longer cooking, but some, such as the sand eel, are not thick and will cook within a minute or two. Of course, the method works best when all the fish are about the same size. Larger fish, such as 6-inch smelt, can be filleted and cooked along with true whitebait. Epicures will point out that whitebait, to be ideal, must be kept alive until the time of cooking, which would limit fresh consumption to towns near the waters from which the fish were taken.

Note that edible minnows should not necessarily be considered the same as whitebait. Although the word came from the Anglo-Saxon *myne*, meaning small, some fish (such as carp) that are classified scientifically as minnows—that is, in the family Cyprinidae—grow surprisingly large. The squawfish—also an edible minnow, said to be delicious when smoked—lives in some of the streams of Washington and Oregon. The northern squawfish (*Ptychocheilus oregonesis*) grows up to 80 pounds.

(See also FISH; GALAXIIDS.)

wombats An old myth in Tasmania has it that the wombat, along with the enchidna and the kangaroo rat, pestered and threw stones at some people during their sleep. After this went on for some time, the men jumped up, caught the animals, and thrust them into holes in the ground. Further, it was decreed that the people had to eat these animals in order to keep their numbers down. That's why, it is said, the wombat, the enchidna, and the kangroo rat live in holes—and why people eat them.

The facts are that wombats do live in holes—and are still eaten even today. There are three extant species in southeast Australia and Tasmania, and the largest one, the common wombat (*Vombatus ursinus*), weighs up to 86 pounds. Wombats have large chisel-like teeth and resemble rodents, especially the woodchucks or marmots, but they are actually members of the marsupial family. Completely terrestrial, they feed at night on grass, roots, shoots, and other vegetable matter. During the day they sleep in their burrows, which may be 100 feet long with several entrances, side tunnels, and chambers. Although all species are more or less protected, the common wombat is considered to be a nuisance in eastern Australia because of its burrows.

(See also MARSUPIALS.)

woodcocks These small, long-billed game birds are widely hunted by sportsmen and are highly esteemed by epicures. They are eaten throughout their extensive range, and are especially appreciated as gourmet table fare in France and other parts of Europe. Not many years ago, the woodcock was often "hung" until it was "high" before it was cooked and eaten. Some experts believed that the bird ought to be hung by the feet, and others thought it should be hung by the neck until it fell from its own body weight. Further, it was hung with all its innards except the gizzard. Some people cooked and ate the innards along with the rest of the bird. In fact, the trail (intestines) was considered by some gourmets to be the best part of the bird.

In time, such practice was frowned on by governmental health authorities. Some restaurants refused to serve high birds, and slowly chefs and hunters quit hanging the birds, or at least didn't let them get "high" enough to drop by their own weight. Hanging was practiced to a limited degree in early America, but was never widespread.

Like most migratory birds, the woodcock has dark meat. Several species grow around the world, with the European woodcock being the largest (up to 14 inches in length). The European bird is also widely traveled, breeding from Ireland across Asia to Japan, and wintering south in China and North Africa. It has even been found in Virginia. The American woodcock, which grows up to 11 inches long, migrates from Nova Scotia to Florida. It can be found westward to Colorado.

All the woodcocks feed extensively on earthworms and have long, flexible bills designed to probe in the ground—preferably soft, wet bottomlands or mud. Their habitat is not restricted to lowlands, however. They also live in wet uplands—and even in mountains. Nevertheless, the woodcook, apparently,

Woodcock

started out as a shore bird, as *Outdoor Highlights*, published by the Illinois Department of Conservation, reports:

> The American woodcock truly is a wonder of nature as a shorebird that has evolved to live in the woods. To adapt to wooded habitat, the woodcock's eyes have "migrated" far back in its head, enabling it to see a full 360 degrees. In the process, the bird's brain flip-flopped to an upsidedown position. The bird also evolved short, rounded wings enabling it to fly through dense cover.

In addition to the American woodcock (*Philohela minor*) and the European woodcock (*Scolopax rusticola*), there are other species, which live in the mountains of Sumatra, Java, and New Guinea. Still other species can be found on the Moluccas islands.

(See also BIRDS.)

worms See GRUBS AND WORMS.

Y

yaks (*Bos grunniens*) The wild yak, an ox, still roams the arid plateau of Tibet. It's a big animal, reaching a height of 6 feet at the shoulder and weighing in at 1,800 pounds. In the past, the wild yak was quite plentiful and was hunted for meat, sport, leather, and its shaggy hair. Although it is quite wary and has a keen sense of smell, the wild yak has declined in numbers in recent years and is now considered an endangered species. The yak has been domesticated in Tibet, and many of the animals are half-tame. Hybrids have been bred as well. Usually, the domesticated yaks are not as large as the wild ones and the colors are different. The purebred wild yak is solid black, whereas the others are usually a mixture of black and white.

Whether wild or domesticated, the yak is really a staple in Tibet, providing both meat and milk. Its tail is highly esteemed. The yak is also used as a pack animal at the higher altitudes, but it is unsuited for tilling the land at lower altitudes where most of Tibet's agriculture takes place. The hide of the animal is also used, and the nomadic herdsmen of the region make tents of woven yak hair.

Part Two

Edible Plants

The plants covered in this section are listed alphabetically under
their common names. Often, a plant has more than one common
name, and this causes some problems of organization. Anyone looking for information on a particular plant that is not covered under a
headword should always consult the index. Because there are so
many common names, it is best to look for the scientific name in
the index if it is known. The current scientific name is usually listed
after the headword at the beginning of the entry, or sometimes
within the entry text, as well as in the index. In a few cases, an entry
may not deal with a specific plant, but is a more general heading.
Such entries include "cacti," which is a brief discussion of a group
of plants, whereas "prickly pears" heads a discussion of a particular
kind of cactus.

A

abutilons More than 100 species of these tropical plants grow in suitable climates and in the greenhouses of Europe and other places. In Brazil, the flowers of *Abutilon esculentum* (called *bencao de dios*) are cooked with meat. The leaves of other species are eaten like spinach in the West Indies and in Asia. The species *Abutilon indicum* is also eaten in India.

acacias Several kinds of acacias provide useful products for man. Gum arabic, for example, is derived from *Acacia genegal*, which grows wild in tropical Africa. Other species provide unusual woods. The koa of the Hawaiian Islands, *Acacia koa*, for example, is used for making ukeleles. Other species are important commercial sources of tannin, and the bark of some trees are used for making rope and twine.

In ancient Egypt, according to Herodotus, cargo boats were made from the acacia tree and were fitted with sails of papyrus. Probably no place on earth, however, relies as heavily on the acacia as the Somali Republic of East Africa, where the Galol tree (*Acacia bussei*) grows. These nomadic peoples use the tree's long roots to make the framework of their huts. The wood provides fuel and charcoal for fuel; the bark is used for making rope and mats; the leaves provide food for livestock; and the fruit, when ripe, is eaten.

acorns If the truth be told, the real staff of life for mankind as a whole has probably been neither wheat nor rice, but instead the wild acorns from white oak trees (*Quercus alba*) and other species of oaks that grow in America, Europe, Asia, and Africa. The white oak (and its close cousins) was popular for food because it has relatively sweet acorns that can be shelled and eaten raw, or roasted like chestnuts. Most (but not all) of the other acorns are more or less bitter, having a high tannin content. This tannin can be leached out, however, so that most acorns can be used in one way or other. Usually, acorns are boiled in one or more changes of water, then they are mashed into a meal or flour and dried. When properly prepared, acorn bread can be quite tasty as well as filling and wholesome. Of course, acorn meal was also used in soups and stews, and roasted acorns have even been used as a substitute for coffee.

Acorns were eaten by early man in Europe, and were a food important to most American Indians. In fact, some tribes considered the better acorn trees to be personal property. Interestingly, certain Indians mixed clay with acorn meal (at a ratio of about 1 part clay to 20 parts meal by weight). The clay was said to make the resulting bread sweet, and to make it "rise" like yeast. A number of other peoples in the world have also eaten clay, or dirt, in one form or another. Such a practice is called geophagy.

Acorns are important food for deer, wild turkey, squirrels, wood ducks, and other animals, and have been used extensively to feed domestic hogs. There are about 300 species of oaks, and some have been highly valued as wood and as a source of tannin and dye pigments. In addition, the bark of two Mediterranean species provide cork.

agaves Often called the American aloe or the maguey, the agave is often considered to be a form of cactus, which it is

---

Seminole Stew

The acorn meal specified for this recipe is assumed to be sweet, or at least not bitter. If the acorns themselves are sweet, they can be ground into a meal or zapped in a modern food processor. Bitter acorns, however, contain lots of tannin, and should be boiled in water before being ground into meal.

> 2 pounds (907 grams) venison
> 2 large onions
> 1 cup (227 grams) acorn meal
> water
> salt and dried red pepper flakes to taste

Cut the venison into cubes about an inch (2.54 centimeters) thick and chop the onions. Put the venison and onions into a Dutch oven, or a suitable pot, and cover with water. Add salt and pepper to taste. (Remember that dried red peppers are *hot.*) Bring to a boil, reduce heat, and simmer for an hour or longer, until the meat is tender. Remove the meat from the pot with a slotted spoon and put it into a serving bowl. Measure the liquid in the pot and retain 3 cups (708 milliliters) of broth. Add water to make to make 3 cups (708 milliliters) if necessary. Bring the broth to a boil, stir in the acorn meal, reduce heat, and simmer for a few minutes, stirring. Pour the broth over the meat and serve.

---

not. It is also confused with the yucca plant, which it greatly resembles. However, the agave tends to be much larger than the yucca, and one species, the common century plant (*Agave americana*), has a flower stalk that can grow as high as 40 feet. Some species even have leaves that grow as long as 8 feet. In any case, agaves comprise about 300 species, all native to the warmer parts of the New World, especially Central Mexico, where the plant is usually known as *mezcal.* Some species of the plant have been cultivated in other parts of the world, and have naturalized in suitable climates (warm and dry).

All agaves have edible buds at the base of the flower stalk, and some of these buds grow up 2 feet in diameter. They are usually baked in an oven (or earth pit) or roasted in the embers of a fire. They are best when baked for a long time at low temperature, after which they have a sweet taste. The Apaches made extensive use of the agave as food, and even pounded the baked buds, dried them, and stored them for future use. Both the Aztecs and the Mayas also used the agave for food and other purposes. The Aztecs made a form of paper, much like the papyrus of Egypt, from the long flower stalks. Roofing was also made from the stalks, a kind of soap was made from agave pulp, and needles were made from the spikes on the leaves. Even today the leaves are still used in parts of Mexico for making the pit-cooked specialty called *barbacoa*, and the leaves impart a distinctive flavor to the meat and vegetables.

In modern times, various species of agave have been cultivated for their fiber, which is used in making floor mats, ropes, twine, and so forth. For example,

Agave

fermenting it for a period of time. *Pulque* was important not only as a beverage but also as a food, and it is even used in some traditional Mexican recipes. "Salsa Borracha" (drunken sauce), for instance, requires *pulque* as an ingredient.

Although *pulque* has been very important in the past, and is still made in Mexico, Cuba, and no doubt other places, it is slowly being replaced by beer made from grains. Remember, however, that *mescal* or tequila is distilled from a form of *pulque* or fermented agave juice. Since tequila is exported to many parts of the modern world, most experts consider it—not *pulque*—to be the national drink of Mexico. Often, a large worm that grows in the agave plant is bottled with tequila and is sometimes eaten. In fact, the agave worm (*meocuilin*) or maguey slug is considered to be a delicacy in some gourmet circles—a reputation that it has held since the times of the Aztecs, who ate it with guacamole.

Some forms of the agave do not bloom frequently—often not for decades—a fact that gave rise to the name "centuryplant." When the plants do bloom, however, they can be spectacular. Some stalks have as many as 300 flowers in a single cluster.

(See also ROOT AND TUBER VEGETABLES.)

sisal (*Agave sisalana*) a native of the Yucatán, has been cultivated extensively in Africa and Indonesia.

Perhaps the most important use of the agave was in making *pulque*, a sort of beer, that, in its weaker stages, also served rather like a soft drink for the peasants. In its more primitive form, it was made by boring a hole into the agave plant, collecting the sap that ran out, and

air potatoes (*Dioscorea bulbifera*) These plants of the tropics are really yams, not potatoes. Like most other yams (there are said to be 700 species of *Dioscorea*), the air potato grows an underground tuber, which in this case is edible but not very tasty. The plant's name comes from the fact that it also grows edible bulbils in the air. These are usually about the size of a man's thumb, and they have received high culinary marks. Air potatoes are usually boiled like ordinary potatoes.

A native of Africa, this species is cultivated in not only in the tropical regions of Africa but also in parts of Asia and the Caribbean. Actually, there are several species of edible "air potato," and many grow wild. According to *The U.S. Armed Forces Survival Manual*, they are especially plentiful in Southeast Asia. This book also says that some of them, like some other yams, are poisonous if eaten fresh.

(See also ROOT AND TUBER VEGETABLES; YAMS.)

akees (ackees) Sometimes called akee apple trees, these fruit trees of Guinea were taken from West Africa to Jamaica by Captain Bligh. Hence, their scientific name, *Blighia sapida*, whose latter part comes from the fact that they belong to the soapberry family, Sapindaceae. The trees are evergreens that bear red fruits with bright black seeds.

Ripe akee fruits can be eaten fresh, but underripe or overripe fruits are disagreeable to some people and may even be poisonous. The fruits are often cooked in Jamaican recipes; in fact, salt fish and akee is almost a national Jamaican dish. The fruits are available canned as well as fresh in Jamaica, and are exported in the can. The edible part of the akee is sometimes called "vegetable brains" because, when cooked, the mass looks like scrambled eggs and brains.

(See also TROPICAL FRUITS.)

alexanders (*Smyrnium olusatrum*) Sometimes called horse parsley, these stalk plants of the Mediterranean region were at one time cultivated as vegetables. They now grow wild. Waverley Root says they were replaced in the 18th century by celery.

alfalfa This popular fodder (*Medicago sativa*) may be the first plant that was cultivated by man. Because its roots grow as deep as 50 feet, it is quite tolerant of droughts, although its yield isn't great in dry weather. The plant probably originated in the Near East, and it still grows wild in southern Russia.

Both the English and the Spanish brought the plant to America, where its seeds were used for bread and its foliage was cooked as a green vegetable. The leaves, high in vitamins A and D, have been used extensively in the manufacture of Pablum (a baby cereal) as well as being used in vitamin tablets.

algae See DULSES; LAVERS.

alpine bistorts (*Polygonum viviparum*) Sometimes called alpine smartweeds, these members of the buckwheat family grow in the rocky soils and slopes in the northern parts of Canada and Alaska, and follows the Rocky Mountains down almost to Mexico. The alpine bistort puts out a spike, which is surrounded in summer with white or pink flowers and red-to-purple bulbils. The bulbils can be stripped off the stalk and eaten raw, either as a trail nibble or in a tossed salad. The young leaves of the plant are also edible, either green in a salad or cooked as a potherb.

The alpine bistort has a somewhat contorted root (the name bistort is from the Latin, meaning twice-twisted) that is nonetheless edible, either raw, boiled, or roasted. The roots receive high culinary marks, at least for a wild root, and the young ones are tender as well as tasty.

Several other bistorts also have edible roots, which in times of need have been ground into flour and used in breadstuffs. They also have edible leaves—but not bulbils. In fact, the *Polygonum* genus includes several edible plants, such as Japanese knotweed and Alaskan knotweed.

amaranths (*Amaranthus retroflexus*) These North American weeds, which are probably of tropical origin, now grow from the Atlantic to the Pacific and up to the edge of Canada. Sometimes called "pigweeds," redroots, or wild beets, the plants are highly regarded by foragers. The young leaves, picked in the spring, are mild and can be eaten raw or boiled. The dried leaves can be used in soups.

The green amaranth grows to a height of about 6 feet, and its head contains thousands of tiny black seeds. These are eaten whole or ground into a flour. In the past, amaranth flour has been widely used and is even available commercially today. The flavor of the seeds (and of the flour) is improved by parching them slightly. At one time, the Indians of the American Southwest cultivated amaranth plants for their seeds.

The green amaranth has a bright red root, which explains the name "wild beet," and which helps in its positive identification. Several other species are also edible.

Be warned that amaranth should be eaten only in moderation when taken from fields that have been fertilized with nitrates, which can be concentrated in the plant.

angelica (*Angelica archangelica*) This aromatic herb of northern Eurasian origin was not widely known in southern Europe until fairly late in human history. It was, however, widely eaten in Iceland and Lapland, where most green vegetables don't grow. The plant has a celerylike leafstalk that can be cooked or eaten raw. The green stems are also candied, bits of which are often used in fruitcakes. From the seeds and roots come oils that are widely used in perfumes and in such liqueurs as Chartreuse.

Apparently, the Vikings brought the plant to southern Europe during the Middle Ages. It was believed to ward off the plague and to protect against poison and mad dogs. Hence, its angelic name. It has also been called the "root of the Holy Ghost."

(See also HERBS AND SPICES.)

apples and pears These great fruits were raised 4,000 years ago in the orchards of the Hittites and they have been very popular in the temperate regions of the world. The fruits are marketed in fresh, dried, and canned form and, of course, the juice is also consumed in one way or another. Modern New Englanders who know all about apple cider should also know that not many years ago the juice of the pear was used in making an alocholic beverage called perry. This drink is still enjoyed in the north of Europe.

(See also CRAB APPLES.)

arrowheads (genus *Sagittaria*) Also called duck potato and wapato, about 40 species of arrowhead grow in the temperate and tropical regions of the world. Most of these have large leaves that are shaped like arrowheads or arrow points, and others are lance-shaped. Some species have tubers that grow from pea-size up to an inch or two in diameter. All of

Arrowhead

the tubers are edible, and the larger ones are sometimes harvested in America, Europe, and Asia. In China, one species is cultivated as a vegetable. (On the West Coast of the United States, the Chinese Americans are fond of arrowhead tubers, which are sometimes seen in oriental markets under the name tule potatoes. In Florida, they are sometimes called swamp potatoes.) Most of the American

Indian tribes made good use of the arrowhead, and the Chinook of Oregon actually traded the tubers in commerce. The tubers are usually gathered as needed and eaten fresh, but can also be dried for future use.

The plant thrives in shallow water over a mud buttom, and the Indians harvested the tubers by wading barefoot along a canoe and dislodging the tubers with their toes. Once free, the tubers floated to the surface. The modern forager might prefer to use a hoe or a four-tine potato fork for this purpose, but should remember that merely pulling up the plant's stalk will not work. Although the tubers can be eaten raw, they are much better when they are boiled like a new potato, which they resemble in flavor and texture.

The arrowhead tuber can be harvested and eaten all year-round, which makes it a good survival food. The plants are also very common and widespread.

arrowroot Originally, the term arrowroot referred to a white starch of high quality that was obtained from the rootstocks of a South American plant with the scientific name of *Maranta arundinacea* and, to a lesser extent, of a very similar *Maranta* plant of the same region. The plants are probably native to Guiana and Brazil, but they gained in the West Indies long ago and are now grown in most tropical countries. It is cultivated in the West Indies, Brazil, Southeast Asia, Africa, and Australia. The starch of arrowroot, important in the cuisine of the West Indies, is used in most parts of the world to thicken stews and gravies, and in sweet dishes and milk puddings.

Similar starch is also obtained from the roots of other plants, and is commonly called arrowroot today. Often, these other forms of arrowroot are also from the tropics. Starch from the cassava plant (*Manihot esculenta*) is processed and marketed in Brazil under the name Brazilian arrowroot. At one time, Portland arrowroot was obtained from a species that grew on the isle of Portland, England. Tacca arrowroot comes from the *Tacca pinnatifida* or pia plant of the South Pacific islands. Other kinds of "arrowroot" come from the roots of plants in India and no doubt other places and other roots. Even starch from potatoes has been marketed as arrowroot. Many peoples have extracted a similar starch from the roots of cattails, water lilies, and other plants.

In France and on some of the islands of the West Indies, a high quality arrowroot, called tous-les-mois, is obtained from several species of tropical canna plants, especially *Canna edulis*. These plants are often cultivated for their beautiful flowers as well as for starch. The flowers of some canna species have many seeds that are quite hard. In fact, the seeds were used by Indians in blowguns. This led to the popular name indianshots, which is applied generally to the cannas group. The Indians of South America also ate the rootstocks of the plants.

(See also ROOT AND TUBER VEGETABLES.)

asparagus About 150 species of asparagus are native to Africa and Europe, on into Siberia. Some of these are eaten, and others are cultivated mostly for ornamental purposes. By far the most significant to man is the common garden species, *Asparagus officinalis*, which is now cultivated for food in America, Europe, and other parts of the world. As it

happens, *Asparagus officinalis* is also the most important wild species, since its seeds are spread by birds from gardens to fencerows and other likely growing areas. The plant was first cultivated in the United States in the 18th century, and it now grows wild, here and there, from coast to coast.

In the wild, the so-called green asparagus is usually the only form that is available, and its shoots tend to be smaller than those of the cultivated form. The green asparagus is also widely eaten as a cultivated vegetable, but another form is also raised. Called white asparagus, it is produced by planting the roots deeper into the dirt. It is cut when only the tip of the shoot sticks out of the ground, so that most of the edible shoot is below the surface. Whether they come from the wild or from the garden, only the young shoots should be eaten. The older stalks and the rest of the plant are more or less toxic.

Nobody knows when mankind started eating asparagus from the wild, but it has been cultivated for food since the ancient Egyptians and it was highly prized by the Roman epicures. In England and elsewhere, the asparagus has been given aphrodisiac status at one time or another.

aspens This genus of popular trees (*Populus*) includes the common aspen of Europe, which grows in Siberia as far north as the Arctic Circle, and the trembling aspen of North America (*Populus tremuloides*), which is also called American aspen, quaking aspen, and quiverleaf. Growing across Canada from Newfoundland to Alaska—and along the Rocky Mountains all the way down to Mexico— the trembling aspen is popular as food for the beaver. Its inner bark is eaten by the snowshoe hare and by man. The bark is a good survival food in parts of the north, and was used as food on a regular basis by some of the American Indians who, in addition to eating the inner bark, also made a syrup from it. Various parts of the tree were also used for medicinal purposes, and its bark is known to contain salicin, the main ingredient in aspirin. (Willow also contains salicin, and both trees are members of the willow family.)

The wood of aspen is both light and tough, and was at one time used for making arrows. It has also been used for making charcoal, which in turn was used in the manufacture of gunpowder. The wood is also used for making matches.

(See also WILLOWS.)

avocados (*Persea americana*) Native to Middle America, growing from Mexico south to the Andes, avocados were enjoyed in native cuisine long before the Spanish arrived. Because they can be grown only in a more or less tropical climate (including south Florida and California) and because they have a relatively short shelf life, they didn't become popular in Europe and in most parts of the United States until after World War II.

There may be an ulterior reason for the increased demand. The word "avocado" came from an Aztec word meaning "testicle tree," a term that probably came about because of the large seed inside the fruit. According to Waverley Root in *Food*, commercial avocado growers, unable to create a demand for their product, took the advice of a public relations expert and flatly denied rumors that avocados were an aphrodisiac. Sales, Root said, immediately mounted.

In any case, the avocado is widely available today and is grown in South Africa, Hawaii, Israel, and Australia as well as in the Americas and the Caribbean.

B

baels (*Aegle marmelos*) These fruit trees are found in the wild and under cultivation in most parts of India and in Burma. Their fruit, also called bael or Bengal quince, is kin to citrus fruit and rich in vitamin C. Usually, the fruit has a yellowish gray rind and an orange pulp, with quite a few seeds. Rather perishable, the bael fruit doesn't lend itself well to supermarket distribution.

(See also CITRUS FRUITS.)

bamboos A number of species of bamboo—members of the Bambusoideae family of grasses—grow wild in the tropical and temperate zones of the world. Bamboo is especially abundant in Southeast Asia. The smaller species grow only to about 6 inches in height, but the larger ones tower 120 feet and are 8 inches in diameter.

All over the Orient the young shoots of bamboo are eaten as a vegetable. For best results, the shoots are harvested before they break the ground. Fresh and canned bamboo shoots are widely available in oriental markets, and sometimes even dried bamboo shoots can be found. In Japan, the shoots are often pickled in vinegar. Some species of bamboo, such as the mosa of China (*Phyllostachus pubescens*) are cultivated in the Orient as well as in Europe and the United States.

Some bamboo shoots contain cyanogens and should be eaten only after they have been boiled. The survivalist should know that the roots of bamboo are also edible, as long as they are peeled and boiled. Also, drinking water can sometimes be found in hollow bamboo stems.

Some species of bamboo have seeds that can be boiled and eaten like rice.

In some areas, the leaves of bamboo are used to feed beef cattle either green as forage or dried as fodder. The tough, hard, resilient stems are widely used for fishing poles, fencing, and for building low-cost houses. In the recent past, some cultures have depended on bamboo, and hundreds of objects were made from the material. In Laos, the Yümbri people—who live in the big bend area of the Mekong River where bamboo is especially plentiful—make sharp knives from fire-hardened split bamboo. Also, in Laos natives cook rice in hollow sections of bamboo stems. This is accomplished by sealing the hole in one end with mud and grass, then adding water and rice and sealing the other end. The sealed tube is placed directly into the fire.

Apart from its sustenance value, the bamboo is important in some places for the manufacture of paper and other commercial products. Even diesel fuel can be made from the bamboo plant.

bananas and plantains A number of varieties of wild and cultivated bananas grow throughout the tropical and subtropical regions of the world. The ordinary yellow banana found in supermarkets (*Musa sapientum*) is one of the most widely eaten fruits, for several reasons. For one thing, it tastes good. For another, it is highly nutritious. It can be picked green and allowed to ripen in transit. It can be cultivated successfully, and it is available throughout the year. The banana is an important cash crop in Formosa, Central America (including many of the Carribean islands), South America, and Africa. Ripe bananas can also be dried successfully, which adds to the fruit's versatility.

Usually, the ripe banana is eaten raw. But it can also be cooked, either ripe or green. Some forms of bananas, notably the plantain (*Musa paradisiaca*) are normally cooked before they are eaten. Plantains are popular in the Caribbean and in Africa, but are not exported in large numbers. Being rather starchy, the plantain doesn't taste as sweet as the banana, and is more a vegetable than a fruit, at least for culinary purposes.

Although a number of kinds of banana plants are to be found in the wild, the fruits are not always available simply because birds, insects, and monkeys are also fond of them. On the other hand, all wild bananas are edible, and unripened fruits can be cooked (boiled, fried, or roasted). In addition to the fruit, the flower buds, the young shoots, and the core or heart of the tree can be eaten either cooked or raw.

The large leaves of bananas are often used to wrap meat prior to cooking (as in the famous *luau* of Hawaii). The tough leaves also have other uses, such as roof thatching. Finally, the survivalist should know that the banana tree can be a source of drinking water; when a banana tree is cut off near the ground, the dished-out stump will fill with water that rises from the roots.

The popularity of the banana is not a modern fad. There are several Sanskrit names for the banana, and the fruit was praised by the ancient Greeks, Romans, and Arabs. Primitive man may have cultivated the banana in Asia. In modern times, one variety has been cultivated especially for making beer.

(See also TROPICAL FRUITS.)

baobabs (*Adansonia digitata*) Native to tropical Africa, these members of the Bombacaceae family grow mainly in the

Baobab

bush country. They attain heights of 60 feet, with enormous trunks up to 30 feet wide. The trunks are so large, in fact, that they are sometimes hollowed out and used as human dwellings. The wood is light and soft, but the bark contains a tough fiber, which is used locally in rope and cloth. A related species, *Adansonia gregorii*, grows in Australia.

The baobab bears a large edible gourdlike fruit, which is often called monkey bread. The pulp of this fruit, in addition to being eaten, is often made into a refreshing drink, which is quite popular in some areas. The leaves of the baobab are also edible, and are usually added to soups. In Africa, dried and powdered baobab leaves are called *lalo*.

barberries (*Berberis vulgaris*) These common plants of Eurasia have been naturalized in Great Britain and the eastern United States, where they grow wild along fencerows, fields, and roadways. The fruits are edible in autumn, when they turn a deep red color. In addition to being used in jelly, the berries are sometimes eaten out of hand, and the tart juice is sometimes used in drinks. Several related edible species grow in America, and in Europe the barberry is sometimes candied. The juice is used to make a syrup and even wine. In the Caucasus, the berries—called *barbaris*—are used to make jams and jellies and accompaniments for roast meats; the dried berries are powdered for use as seasoning for dishes such as *lyulya kebab*.

bark The bark of a tree does not receive high culinary marks, but it can be a valuable survival food. Some American Indians ate it on a regular basis, and often ground it into flour for making bread. Most barks are edible but some are better than others. Aspen, birch, willow, and pine bark is comparatively good and, in general, those trees that are gnawed by the beaver will have palpable bark. As a rule, the inner bark—next to the wood—is by far the best part. The outer bark often has large doses of bitter tannin.

beans See GAO BEANS; JÍCAMAS.

bearberries (*Arctostaphylos uva-ursi*) These plants grow in the colder but subarctic regions of Europe, Asia, and North America. They also dip well down into the United States, especially along the Rocky Mountain range. They are evergreen shrubs that grow in dry, rocky soil. As their name implies, their berries are a favorite food for wild bears. Grouse also like the berries, and deer browse on the leaves.

Man also eats the small, dull red berries. They ripen in summer and fall, but tend to cling to the bush throughout the winter, which makes them a good survival food. When eaten raw, the berries are rather tasteless, but cooking brings out their flavor. They are often used in pies, and are sometimes mixed with other berries. Some people use the leaves to make a tea, and the American Indians believed that bearberry tea helped cure (or prevent) urinary problems. Also, the bearberry's leaves can be dried and smoked. Sometimes they are mixed with tobacco. The American Indians called such a mixture *kinnikinick*, a name that is still used for the plant in some areas.

The alpine bearberry (*Arctostaphylos alpina*), a related species, yields a black berry, which is also edible.

bedstraws The common name for several small plants with flowering stalks of the *Galium* genus of the madder family, bedstraws were at one time actually used for making beds. Some writers believe that yellow bedstraw (*Galium verum*) was the hay used in the manger at Bethlehem where Christ was born. In ancient Greece, it was used in cheese making, owing to the fact that the leaves and stems act as a curdling agent. Also, the plant has been used in England to make a yellow dye to color cheese and women's hair. For these reasons, this species is also called Our Lady's bedstraw, maid's hair, and cheese rennet.

Another common species (*Galium aparine*), called cleavers, bedstraw, catchweed, and goose grass, grows over most of the United States as well as Eurasia. When young, the leaves of this species are sometimes eaten as a potherb. The plant bears small seeds, which can be roasted and used as a coffee substitute. In fact, cleavers is a relative of coffee. High in vitamin C, the plant was once of value as a cure for scurvy and has been widely used in America as a spring tonic.

The northern bedstraw (*Galium boreale*) grows across the northern states, Canada, and Alaska. About 60 other species of bedstraw grow in North America, and none are known to be poisonous.

beech trees Kin to the chestnuts and oaks, the beech trees include two important species. The American beech, *Fagus grandifolia*, grows from New Brunswick to Florida, west to Texas and up to Minnesota. The European beech, *Fagus sylvatica*, grows from the British Isles and Norway down to the Mediterranean and across parts of the U.S.S.R. and Asia Minor. Both species have edible nuts. In fact, the Latin word *fagus* comes from a Greek word meaning "to eat." In England the beechnut, or beechmast, was at one time called "buck," and the county Buckingham once had good forests of beech trees.

In America, the nuts were important to the Iroquois and other Indian tribes, who ate the nuts fresh or stored them for future use. They even ate the immature nuts as well as the young leaves in spring. The inner bark of the tree was eaten in times of need. The early settlers also made good use of beechnuts. In addition to eating the nuts, they roasted and ground them for a coffee substitute. In some areas, the nuts were even used for hog feed.

Beechnuts were also widely used as food in Europe. Beech bark was ground into flour in times of famine. Beechnuts were used to feed swine, and in France the nuts were used to fatten domestic poultry. In the past, the French also used beechnut oil for cooking.

Partly because the beechnut has been been so abundant in the past, it may well have been an significant part of the diet of early man in Europe. The nut is 22 percent protein.

(See also NUTS.)

berries   See BARBERRIES; BEARBERRIES; BUFFALOBERRIES; CRANBERRIES AND BLUEBERRIES; CURRANTS AND GOOSEBERRIES; ELDERS; HACKBERRIES; HAWTHORNS AND HAWS; HIGHBUSH CRANBERRIES; JUNEBERRIES; MANZANITAS; MOUNTAIN ASHES; MULBERRIES; SNOWBERRIES; STAGHORN SUMAC; TIS; WILD STRAWBERRIES.

birch trees Some 40 species of birch trees grow in the cooler regions of the Northern Hemisphere. In North America, the most useful of these are the sweet birch (or black birch), *Betula lenta*, and the yellow birch, *Betula alleghaniensis*, both of the Northeast and Appalachia. These trees provide a sap, like maple, that can be made into syrup or beverages, both soft and more spirited. This copious sap, just as it comes from the tree, is low in sugar content and can be used for drinking and in cooking as a substitute for water. The sap of birch doesn't contain as much sugar as that of maple, so

Black birch

Easy Birch Beer

Tap trees when sap is rising. Jug sap and throw in a handful of shelled corn. Nature finishes the job.

———————————————

—Foxfire 2

that more of it is needed to make syrup. When properly made, birch syrup tastes somewhat like sorghum.

The inner bark of birch has often been used as a survival food. It can be eaten raw, shredded and cooked like noodles, or pulverized and used in soups and breadstuffs. The inner bark is also used to make tea. But modern man should remember that stripping the bark off the trees can kill or disfigure them.

The young twigs can also be used to make a tea, but they should be merely steeped in hot water, not boiled. Boiling destroys the wintergreen flavor and, it should be added, the birch tree—not the wintergreen plant—was at one time the source of the oil of wintergreen sold in commerce. For the best results, the tea can be made with fresh birch sap instead of water, and the tea can be consumed hot or iced. Tea can also be made from dried twigs or bark, but it should be stored properly in a cool place to preserve the wintergreen flavor.

The bark of the paper birch (*Betula papyrifera*) was widely used by the Indians as an outer skin for their canoes. The white birch (*Betula pendula*) grows widely in Europe, Asia Minor, and Siberia. It is an important timber tree. Its bark, which is more or less impervious to water, has been used in liquid containers and as a local roofing material.

Of all the species, the sweet birch is probably the best for sap and survival food. But the rest are edible—or at least aren't poisonous.

bitterroot (*Lewisia rediviva*) This regional North American plant grows in the mountains from British Columbia south to Arizona. It's a short plant with a few leaves and a short stem on which, in spring, appears a beautiful red flower about 2 inches in diameter. The bitterroot often grows in rocky soil on ridges, and sometimes in great profusion in high mountain valleys. The plant was collected by the Lewis and Clark expedition, and was later given the botanical genus *Lewisia* in honor of Meriwether Lewis.

Bitterroot is the state flower of Montana, where both a river and a mountain range are named after it. In spite of the name, the taproot of the plant was an important food for the Indians and, no doubt, for the mountain man.

(See also ROOT AND TUBER VEGETABLES.)

blackberries and raspberries (genus *Rubus*) Three or four hundred species of these berry-bearing vines or shrubs (sometimes called brambles) grow wild in the temperate regions of the world. Most of these have edible fruits. As a rule, the fruits are eaten out of hand or with cream and sugar, cooked in pies, or made into jellies and jams. Wine is also made from the berries. Tea is brewed from the leaves of many species, and the young tips of most species can be eaten in spring. The blackberry and its cousins have also been widely used in folk medicine.

The *Rubus* genus includes blackberries, wineberries, thimbleberries, raspberries, and dewberries. One sort or

---

Mustard Sauce

To make an outstanding sauce from wild mustard seeds, brown some flour in the oven. Then mix equal parts of browned flour and powdered mustard seeds. Stir in a mixture of half water and half vinegar until you get a sauce that spreads just right. Stir in a little salt. Other spices and herbs can be added, such as grated horseradish. Also, the strength of the mustard sauce can be changed considerably by altering the ratio of flour to mustard powder.

---

another grows over much of North America, including the Arctic, where they are sometimes valuable as a survival food. There are also many cultivated varieties, such as the loganberry.

black locusts (*Robinia pseudoacacia*) The more widely known of several species of North American locust trees, the black locust is often called false acacia, as the scientific name might imply. The tree has been introduced in Europe, where it is sometimes cultivated as an ornamental tree. (It is sometimes called acacia in England.) The tree has dense wood, which has value as timber, and as such it is popular in Hungary. (One variety is called shipmast locust.) The black locust also has an adventurous root system that makes it valuable for soil erosion control.

Be warned that some parts of the black locust are poisonous. The flowers and seeds, however, are edible and are sometimes used for food. The flowers, which grow in large clusters, are usually eaten in fritters, and they are also used to brew a tea. The oily seeds are used as beans or peas, and they can be dried like beans for winter use. The seeds have an acid taste, but this disappears after boiling.

black mustards (*Brassica nigra*) A native of Europe and Asia, this common plant has been introduced to America and grows wild almost everywhere, along with several closely related species. Wild mustard often thrives along the edge of fields, pastures, and other clearings— even vacant lots in cities and towns.

The very young, tender leaves make the best eating. These can be used along with lettuce and other greens in a tossed salad, or they can be boiled in a little salted water and served up like ordinary turnip greens or spinach. The older mustard leaves can also be eaten, but they are tougher and stronger of flavor, often with a somewhat bitter taste. Seasoned epicures may relish a slight bitter taste, but beginners would be well advised to boil the older mustard leaves for a few minutes, then drain and change the water.

Mustard greens are quite nutritious, being especially high in vitamins A, B1, B2, and C. In addition to the leaves, the buds of wild mustard can be eaten, like tiny broccoli. These are usually boiled for only two or three minutes and then seasoned with a little butter, salt, and pepper.

After the plant reaches maturity, the seeds can be harvested. It's best to get the whole seed stalks, dry them in the sun for a week, and then frail them over a large bucket or sheet. The seeds can be sprinkled over salads, or they can be ground to a powder and used as seasoning. (See also GREENS.)

blueberries See CRANBERRIES AND BLUE-
BERRIES.

Brazil nuts Well known and tasty, these
large nuts are actually three-sided seeds
with hard shells. The seeds grow inside a
larger fruit, or woody shell, of a tropical
tree (*Bertholletia excelsa*) of South America.
The seeds grow like sections of an orange,
which explains their odd shape. Between 8
to 24 seeds grow inside a single fruit, which
can be up to 8 inches in diameter. In its
native South America, the Brazil nut is
called *juvia*. In France, it is sometimes called the American chestnut.
It is also called paranut, creamnut,
castanea, and butternut. In England and
America, it is a table nut eaten traditionally at Christmastime.

Also called the butternut, the swarri
(of the *Caryocar* genus) of tropical South
America, especially the Guianas, is even
larger than the Brazil nut. It is exported,
on a limited basis, to the United States
and Europe. The swarri is also called the
souarinut, pekea nut, and piki.
(See also NUTS.)

breadfruits (*Artocarpus altilis*) These
plants, whose fruit is a staple in the South
Pacific, were first cultivated in the Malay
archipelago, and from there they spread to
many other islands during prehistoric
times. The breadfruit is not a true fruit, but
more like a large potato or an eggplant.
The fruits usually grow from 4 to 8 inches
in diameter. They are also profilic, so that
a good tree can provide for an entire family
for a year. The breadfruit can be eaten in
various stages of ripeness, and is cooked
accordingly. However, it does not keep
well and in the tropics it is often fermented
before use. (It can be frozen, however.)
Growing to 60 feet high, the tree has large

leaves—up to a foot long—and is used
for more than food. The tree provides
wood for canoes and other purposes.
The fibrous bark was at one time made
into cloth by the islanders. Cuts in the
trunk yielded a sticky milk that was used
for glue and caulking.

The islanders cooked the breadfruit
in pits in the ground, using alternating
layers of hot stones, breadfruit, and
leaves. It is still cooked in this manner.
The breadfruit is often sliced and fried
like potatoes, and it is also baked, boiled,
roasted, broiled, and grilled. When
mashed or puréed, it is used in soups,
pies, and cakes. It can also be eaten raw
if necessary. Reportedly, the seeds of the
breadfruit taste like chestnuts.

News of the breadfruit came to Europe by way of British sailors, and Captain James Cook investigated the plant
on one of his voyages. He said the fruit
tasted none too good, but a botanist with
him (as well as most of the sailors) gave
it better culinary marks. Almost all
things considered, it was decided that it
would be an ideal food for the slaves who
worked the sugarcane fields of the West
Indies. To this end, George III dis-
patched Captain Bligh aboard the
H.M.S. *Bounty* to transport a number of
suitable trees from the South Seas to the
West Indies. Bligh lingered among the
islands until he loaded the *Bounty* with a
thousand prime saplings. On the way to
the West Indies, he depleted the ship's
drinking water in order to keep the
breadfruit trees alive. This led to the
famous mutiny, in which Bligh was set
adrift in a small craft and the breadfruit
trees were thrown overboard. Bligh sur-
vived the seas, and a few years later returned to the islands to complete the
mission.

The breadfruit trees did very well in the West Indies, but the trouble was that the slaves wouldn't eat the fruits. In time, however, the breadfruit became more or less accepted, and is now an important food in the West Indies. It is also cultivated in Mexico and south to Brazil. To a limited extent, breadfruit is also available in some markets in the United States.

(See also JACKFRUITS; TROPICAL FRUITS.)

breadroot (*Psoralea esculenta*) A perennial herb, breadroot grows on the plains and dry, rocky woodlands of mid-America, from Texas to southern Canada. The plant grows only about a foot tall and often goes unnoticed. Breadroot has a large taproot, sometimes branched or clustered, up to 4 inches long. This root was eaten by the plains Indians as well as by the settlers. It can be eaten raw, roasted over a campfire, boiled, or baked. As the name implies, it can also be dried and ground into flour for use in making bread. The tubers can be peeled and dried slowly for use during the rest of the year, usually by reconstituting the dried tuber with water.

A similar edible species, *Psoralea hypogala*, also grows in the same area, but has a smaller root. Both species have to be gathered in summer, when the small top is visible. The top withers and blows away in the strong winds of the plains, making the large tubers difficult to find.

An attempt was once made to raise the breadroot in Europe as a food crop, but for one reason or another the attempt failed. The breadroot was, however, an important plant in early America and saved many people from starvation. It is still a valuable survival food.

Breadroot

(See also ROOT AND TUBER VEGETABLES.)

brooklime (*Veronica americana*) Brooklime grows along the banks of streams and ponds, sometimes together with watercress. Like watercress, brooklime is eaten raw in salads or cooked as a potherb. Generally, it grows all across North America, except in the extreme northern regions, from lower Alaska to Nova Scotia. It isn't widespread in the southeast, but in the west it grows south into Mexico. Similar species of Veronica are also edible, and a European species (*Veronica beccabunga*) grows in Europe, Asia, and North Africa, and has been naturalized in the United States. This species is eaten in Europe where, according to *Larousse Gastronomique*, it is also called water pimpernet and is used like watercress.

buffaloberries (*Shepherdia argentea*) Hearty North American plants, buffaloberries grow in the meadows and along streambeds in the great plains from Manitoba and Saskatchewan south to Kansas and Nevada, as well as in parts of California. They are bushy shrubs that grow up to 16 feet tall. In summer, they bear many clusters of berry-sized, bright red fruits. The shrubs can be found in the wild, and have also been sold by nurseries for hedge plants and fruit bearers. The berries are sometimes called rabbitberries and Nebraska currants.

The tart fruits ripen in summer and take on a sweeter taste in autumn. They can be eaten out of hand, cooked in pies and jellies, or dried for future use. Hot tea as well as cold beverages can be made from fresh or dried fruits. A sauce similar to cranberry sauce is made with the berries and is eaten with buffalo and other good meats.

bugleweed (*Lycopus uniflorus*) According to *Field Guide to North American Edible Wild Plants*, this perennial herb has edible tubers. It grows in the northern part of the United States and southern Canada. The tubers can be dug in spring, autumn, or winter, and they can be eaten raw, boiled, or pickled. Some other species of *Lycopus* have no tubers. (See also ROOT AND TUBER VEGETABLES.)

buglosses (*Anchusa azurea*) The flowers of these plants of southern Europe were once eaten in salads. Believed to be a tonic, bugloss salad was a favorite of Louis XIII. The flowers are still eaten on a limited basis, and the leaves are sometimes cooked like spinach.

Bugloss is in the same family as borage (Boraginaceae), which also has edible leaves and flowers. Both borage and bugloss have been eaten as a remedy for melancholy. Native to the Mediterranean, it still grows wild in the hills of Sicily. The Romans ate borage and believed that it gave them courage.

bull thistles (*Cirsium vulgare*) Native to Europe and Asia, bull thistles have been introduced in North America, where they grow just about everywhere. They are ugly plants with many spines, but they develop a beautiful purple flower. They grow along roadways, fields, and meadows. The leaves, usually gathered with the aid of gloves and knife, are tasty if picked while young. After the spines are removed, the leaves can be eaten in a green salad or cooked as a potherb.

The bull thistle puts out long flower stalks, which grow up to 5 or 6 feet tall. When young, these can be peeled, cut into sections, and boiled in salted water for a few minutes. After draining, they are usually eaten with a little butter. The taste is said to be quite good, but the older stalks tend to be fibrous and tough. In *Wild Edibles of Missouri*, Jan Phillips reports that she eats them by pulling the stems through the teeth, stripping off the good part and discarding the stringy part, as when eating artichoke petals.

The roots of the first-year plants (the bull thistle is biennial) can also be used for food. After they are dug up, they are peeled and boiled until tender. Then they can be sliced and fried. The roots can also be boiled until very tender, dried, and ground into flour for use in breadstuffs.

A number of other edible *Cirsium* species of thistles grow in various parts of the world, and they have been called the knights in armor of the vegetable kingdom. As legend has it, when the Danes invaded Scotland in the 11th century, they sneaked in barefooted for a surprise attack and one of them stepped on a thistle. The battle cry was thus let out, and the thistle became the national emblem of Scotland.

The bull thistle should not be confused with the sow thistle, another species entirely.

bulrushes Because common cattails are often called bulrushes, confusion regarding which is which dates far back into Biblical history. The plant in which the baby Moses was hidden was called bulrush, but it was probably papyrus, which grew along the Nile and was at one time quite important in commerce.

Further, there are several species of bulrush that grow around the world. In North America, two edible kinds grow over most of the United States and Canada. These are the great bulrush, *Scirpus validus*, and the hard-stem or common bulrush *Scirpus acutus.* The great bulrush and the hard-stem bulrush are called tule. Other species include the chairmakers-rush, *Scirpus americanus*, no doubt named because it was used for weaving mats for chair seats. The common bulrush *Scirpus acutus* is probably the most widespread species in North America, and it grows along salt- and freshwater marshes, inlets, lakes, and streams from British Columbia to Arizona, across Florida, and north to Newfoundland. The bulrush stems grow up to 9 feet tall and were used by the American Indians to build boats and mats. In fact, the Indians of the Andes still make boats of bulrush for fishing on Lake Titicaca, a large mountain lake.

According to *The U.S. Armed Forces Survival Manual*, the bulrush grows in North America, Africa, Australia, the East Indies, and Malaya. (The *Manual* doesn't list South America.) Wherever they are found, bulrushes can be identified by the long leafless stem topped with a turf of spikelike flowers. All of the rushes have edible parts, and they are good plants for the survivalist to know as they can provide food of one sort or another throughout the year.

In spring, the young shoots can be eaten raw or cooked like asparagus. The young rootstocks can be sliced and fried like potatoes or boiled. A sweet syrup can be made by first boiling the rootstocks and then boiling the liquid until it thickens. The older rootstocks can be peeled, dried, and pulverized for use as flour in breadstuffs. Pollen to use in breads can be obtained by shaking the flowering heads in a bag or bucket during summer, and in autumn and winter the small, flattened seeds can be obtained in the same manner. The seeds can be ground into flour.

(See also CATTAILS; PAPYRI; REEDS.)

bunchberries (*Cornus canadensis*) Members of the dogwood family, these small, low-growing plants produce beautiful white flowers and clusters of red berries. They grow across Canada, dipping down into the United States here and there, especially in the northwestern states. In summer, the ripe berries are eaten raw or made into pies. Since they are rather tasteless, they are best mixed with tart fruits of other plants.

Several related species are also edible, especially the *Cornus swecia* or Swedish bunchberry, which is sometimes called Lapland cornel. The berries from this plant have more flavor, and Laplanders make a pudding from them. Lapland cornel is truly circumpolar, and in some places it even grows north of the Arctic Circle. Still another species in the dogwood family, *Cornus mas*, produces the cornelian cherry of Europe and Asia, which is often made into preserves or jelly. About the size of an olive, the cherry has a bright red color.

burdocks (*Arctium minus*) Siberian natives, the common burdocks (according to Bradford Angier's *Free for the Eating*) marched across Europe with the Roman legions, crossed the Atlantic with the early settlers, and now grow wild and unwanted over much of North America. They belong to the thistle family. The settlers used various parts of the plants for food, and the Iroquois soon learned how to use them as well. The plants (or perhaps a related species, *Arctium lappa*) are actually cultivated in Japan and other parts of the East.

In *Stalking the Wild Asparagus*, Euell Gibbons said that he first became acquainted with the burdock's culinary possibilities in Hawaii, where Japanese truck farmers grew it for the market under the name *gobo*. In Hawaii, the plant is believed to give one strength and to have aphrodisiac powers. Gibbons also reported that he has seen burdock in Japanese markets in Chicago under the name wild gobo. It is now marketed in other Japanese markets, and is sometimes called beggar's button. The plant is popular in Taiwan as well as in Japan and Hawaii.

In any case, the wild burdock is ideal for the forager because it is easy to identify by its large leaves (up to a foot long) and its sticker seed pods. The roots of the first-year plants can be gathered for food

during the summer, preferably in June or early July. These roots are peeled and are usually boiled in two changes of water. The young leaves, if gathered in early spring, can also be eaten as a vegetable, usually after being boiled in two changes of water. The young leafstalks can be peeled and eaten in a green salad, or cooked like asparagus. The large flower stalk can also be peeled, revealing a pith that can be eaten raw or cooked. When cooked in syrup, this pith also makes a kind of candy.

But from a commercial viewpoint the carrot-shaped root is really the important part of the plant. These can grow up to 2 feet in length, although the root of the wild burdock is usually much smaller. When cooked properly, the root has grayish white meat with a flavor similar to that of artichoke hearts.

butternuts (*Juglans cinerea*) Closely related to walnuts, butternuts grow in the northeastern part of America, from New Brunswick south to Tennessee and west to North Dakota. Of medium size, the trees bear nuts that are about 2 inches in diameter. The nuts are hard to crack, and it's difficult to pick out the meats once the shell is cracked. They are, nevertheless, delicious.

The mature nut meats can be eaten raw or used in baked goods and confections. The meats can be ground into a meal and used in breadstuffs. The ground meats are also mixed with honey to form a spread for toast. When cracked the nuts are boiled (often as an aid to removing the meat from the hull) and a useful flavoring oil can be skimmed off the top and used like butter. The sap, when tapped from the tree in spring, can be boiled down like maple syrup, and can even be made into a sugar. The green or immature nuts, gathered in summer, can be pickled, husk and all.

According to *Earth Medicine-Earth Foods*, A.C. Parker, an anthropologist of Seneca Indian ancestry, says that the Indians made a baby food by mixing dried pulverized butternut (and walnuts) with pulverized venison or bear meat. This mixture was added to warm water and used like milk. Parker writes that "the nursing bottle was a dried and greased beargut. The nipple was a bird's quill around which was tied the gut to give proper size."

(See also NUTS.)

C

cacti A number of cacti grow in various parts of the world. Most of these were originally from arid or at least semiarid regions, but a few are native to tropical places. A good many species now grow widely outside their original range; the prickly pear, for example, was once found only in the New World, and now it grows wild in southern Europe, North Africa, the Middle East, and Australia. The American Indians used about 40 species of cacti for food—and for water. The large barrel cactus was especially valuable as a source of water for the desert Indians and early travelers. They cut off the top, scooped out the pulp, and pounded it to release the water. They also used the hollowed-out plant as a cooking utensil, merely by putting in raw food, covering it with water, and dropping in heated rocks.

The fruits of several cactus plants were widely used as food, and were dried for future use. They were also used to make a fermented beverage. In some cases, as with the prickly pear, the leaves were also eaten. Cacti, of course, still grow wild, but some of the best ones for eating, such as the saguaro, are scarce these days and may be protected by law in some areas. The good news for modern gourmets is that some cactus fruits, often called Indian figs, are now grown commercially and are sometimes available in supermarkets.

Be warned that some cacti, such as peyote, are hallucinogenic, and the hedgehog cactus (*Echinocereus coccineus*) may be poisonous. Some of the American Indians are reported to have called it *Tjeenáyookísih*, meaning plant that would "twist the heart" if eaten.

(See also PRICKLY PEARS; SAGUAROS.)

Cactus Fruit Windfall

The Indians of Lower California [Baja], long extinct, lived in a relatively barren environment where food was scarce, but prickly shrubs were fairly abundant in favored places. Among them is the pitahaya cactus, a member of the family that includes the tall and impressive saguaro of our own Southwest. The sweet pitahaya (there is also a sour variety) bears a fruit that is round, as large as a hen's egg, and covered with a green, prickly shell. Its pulp is white and red, juicy, and fairly sweet. Inside the pulp small black seeds are scattered. The fruit ripens in the middle of June and can be eaten for about eight weeks. During the eighteenth century the Indians used to come to pitahaya clumps, where fruit could be picked in the hundreds. They would feast on it as long as it was available, and would grow fat. Then they would carefully pick the seeds out of their own accumulated excrement, roast them, grind them, and eat them. The Jesuit priest Father Jacob Baegert, who described the practice, called it a second harvest.

—Carleton S. Coon, *The Hunting Peoples*

California laurels (*Umbellulria californica*) Native to California and Oregon, the California laurel, also called California bay and Oregon myrtle, is a medium-to-average sized tree, although some specimens have grown up to 5 feet in diameter. Its wood is prized in cabinetmaking and its leaves are sometimes dried and used in cookery as a substitute for bay leaves. The Indians of the region sprinkled crushed leaves about the lodge to repel fleas. The tree bears an olivelike fruit, but only its seeds were eaten by the Indians of the region. After being removed from the fruit pulp, the seeds were roasted before being eaten. More often than not, the seeds were roasted and then ground into a flour for use in breadstuffs.

camass lilies Sometimes called blue camass, these lilies have edible bulbs. They are usually baked or boiled, or eaten raw. The American Indians ate them in large numbers, and cooked them in a pit in the ground lined with heated stones. Sometimes large batches were cooked on a communal basis. The camass bulb is low in starch and high in sugar, which gives it a sweet taste. After being baked, the bulbs can be sliced and dried for future use.

The eastern species is called *Camassia scilloides* and the western species is *Camassia quamash*. The quamash (an Indian name) was so important to the Indians of the northwest that tribal wars were fought over the control of the rights to collect it from certain areas. The Indians usually gathered the camass only after the plants started blooming. And for good reason: The edible camass has blue flowers and the poisonous death camass has white or yellowish white flowers.

(See also ROOT AND TUBER VEGETABLES.)

cañaigres (*Rumex hymenosepalus*) These plants of the American west have been called wild rhubarb and wild pieplant because their edible stalk can be used as a substitute for the regular garden rhubarb that is commonly used for making pies and preserves. Cañaigre contains large amounts of tannin, which makes it useful for tanning leather (or softening buckskin), and which explains another popular name, tanner's dock. The plant has large roots, which have been used medicinally as an astringent. A tea was once made with the roots and used as a sore-throat gargle. The Navajos also used the roots to make a yellow dye. The plant has also been used to make herbal tea which, according to *Magic and Medicine of Plants* (Reader's Digest Group), has been marketed as being made from wild red American ginseng.

capers (*Capparis spinosa* and similar species) The buds of the caper bush are pickled and sold at high prices. As a rule, the younger the bud, the better it is and the higher the price. The caper is cultivated in the south of France, Sicily, and other places, and the plant grows wild in arid regions of North Africa, Arabia, Turkey, India, and Indonesia. According to *The U.S. Armed Forces Survival Manual*, the fruits as well as the flower buds are edible. Capers are used in salads and other foods.

carambolas (*Averrhoa carambola*) Tart and tasty, these fruits have a very unusual shape, giving them the names star fruits, Chinese star fruits and five-angled fruits. They have a thin, glossy yellowish skin and don't require peeling or seeding. Slicing will do. Some are a little tart, and others are sweet; it's difficult to tell the difference except by taste.

These strange fruits have long been cultivated in Asia, South America, Central America, the Caribbean, and Hawaii. The carambola is now being grown in Florida and is thus more and more available in American supermarkets.

carobs (*Ceratonia siliqua*) Native to the arid lands of the eastern Mediterranean, carobs have been transplanted and now grow across Arabia and on into India. These evergreen trees grow about 50 feet tall and bear small red flowers and edible seedpods. The pods grow as long as 12 inches and may have up to 15 seeds in a sweet pulp. The pods are about 50 percent sugar, and are often used for livestock food as well as for human consumption, especially during times of scarcity. The seeds are also used in the manufacture of carob gum, which has various industrial uses.

The carob tree is also called "locust," and some people believe that John the Baptist ate seeds or pods of the carob instead of the locust insect while wandering in the desert. The tree is sometimes called "St. John's Bread."

cashews (*Anacardium occidentale*) Just about everybody knows the cashew nut, but the cashew apple usually requires explanation. Both come from a tropical or subtropical plant indigenous to Central and South America that grows up to 40 feet high. The fruit (shaped more like a pear than like an apple) forms from yellowish pink flowers; the nut grows on the end of the fruit, almost as if it has been stuck into the end of the apple. In any case, the fruit is eaten in jams and jellies, and is also used in beverages. Remember, however, that the plant is related to poison ivy and poison sumac and will produce the same rashlike effect on skin.

The edible part of the nut is hidden by an outer covering and a hard inner shell, both of which have to be removed. Between the inner and outer shells is a kind of oil, which is quite poisonous and blistering. The oil can be removed by roasting the whole nut; but the resulting smoke or steam is also harmful and can cause temporary or even permanent blindness. (There are, however, safe procedures that can be used for large-scale production.) After the outer covering is safely removed, the hard shell is cracked by hand. The kernels are heated again and the skins are removed. What's left is edible.

The cashew has been introduced to Africa, Asia, and other tropical and subtropical areas, and is widely cultivated. Cashew nuts are an important export for India.

(See also NUTS.)

cassavas The two species of these plants—shrubs that resemble poinsettias—have large irregular roots that look somewhat like a sweet potato. The roots have an ugly barklike skin that covers white, firm flesh. They grow up to 3 feet long and as much as 9 inches in diameter. One kind, the bitter cassava (*Manihot esculenta*), contains the highly poisonous hydrocyanic acid and cannot be safely eaten when fresh. Nonetheless, it is important commercially. The sweet cassava (*Manihot dulcis*) is perfectly safe.

Both plants probably originated in South America, but they have been widely transplanted to the West Indies, Africa, the South Pacific, and the Malay archipelago. They are, in fact, an important staple food in these areas. The plants are also raised in Florida, and have for a long time been eaten in that state by a few local people. Fresh cassava is beginning to appear on American markets, and canned cassava is also available. The plant is often called yuca, yucca, manioc, tapioca, and mandioca.

The cassava is often boiled or baked as a vegetable. It is also fried as chips or used in fritters, especially in the Caribbean and used in soups, desserts, and other dishes. When cooked, the starchy flesh becomes glutinous, and is obviously quite versatile. In South America and in the Caribbean, the juice of the roots is boiled down and used in an important cooking sauce called cassareep, which also contains brown sugar. This important ingredient to native American cooking is available in bottles, and of course many cooks make their own. Cassava roots are also the source of tapioca, a popular thickening starch for soups and stews. Tapioca is a major export for some areas that grow cassava.

The roots are also highly popular in making puddings, especially in the South Pacific, where they are combined with coconut and topped with tropical fruits. In Asia, the roots are grated, shaped into pones, and then baked. The roots are also dried and used in making meal, which is sold commercially and used to make a hard flat bread called cassabe. Of course, the roots are often dried for storage and keep nicely when frozen.

The plant grows wild and is dangerous to those foragers who don't know that it (or at least *Manihot esculenta*) must be cooked. In South America, primitive hunters concentrate the poison and use it to tip arrows and blowgun darts.

(See also ROOT AND TUBER VEGETABLES.)

Cattail Jelly

Believe it or not, the roots of cattail plants can be made into jelly. Here's a recipe from Jan Phillips, author of *Wild Edibles of Missouri*, who says that she makes the jelly after the first "flour" has been rubbed out: The jelly is made by boiling the roots for 10 minutes in enough water to cover them. For every cup (236 milliliters) of liquid, add equal amounts of sugar and a package of pectin per every four cups (944 milliliters) of juice. The jelly is delicious, somewhat resembling honey in both color and taste.

---

cattails These common plants grow along the water's edge and in wet spots all over the world, except in the tundra and the deserts. They can sometimes be found in wet ground, as in ditches, but are usually associated with a marsh, pond, lake, stream, or swamp, in both fresh and brackish water. Sometimes, cattails can be used to locate fresh water. When they grow in a seemingly dry spot, water is usually within a foot or two of the surface.

The two North American species are the common cattail (*Typha latifolia*) and the narrow-leaved cattail (*Typha angustifolia*). Other species live around the world, called by a number of popular names, but they are all pretty much the same. The cattail has been called the "world's supermarket" because it has been used in so many different ways. In early spring, the roots put out new growth that pushes upward to the surface of the ground. These tiny sprouts can be peeled and eaten raw, preferably in a salad, or they can be cooked. The white flesh is mild and crunchy, like a bamboo shoot or water chestnut.

After the shoot breaks through the surface, it quickly becomes a stalk. If this shoot is pulled early enough (usually when less than 3 feet tall), it will break off at the root. It can be peeled, revealing an inner core that can be eaten raw or cooked as a vegetable; this food has been popular in parts of Russia. Some people, especially the British, call it "Cossack asparagus."

Later, bloom spikes will appear, covered with green papery sheaths, similar to corn husks. Under the sheaths, a firm substance grows down to the hard coblike stalks. This substance is made up of densely packed, tiny flower buds. The sheath is removed and the head or ear is boiled in water for a few minutes. Then the flower buds are gnawed off, in a manner much like eating corn on the cob. When picked early enough, the bloom spike is quite good, especially if rolled in melted butter before eating. If it is too old, it will be dry and hard to swallow. The young spikes can also be shucked and scraped, with the bud substance being cooked in a casserole.

After the papery sheath bursts off, the spike will turn a bright yellow as it becomes covered with pollen, which can be gathered by putting a container or plastic bag over the spike and shaking the stalk. It can be used as a yellow flour for baking breads, pancakes, and so on. Most modern foragers mix it half and half with wheat flour.

During summer—and on into fall and winter—the cattail roots can be made into a white flour. The best way to do this is to peel the roots and crush the core in a container of water. Then strain the water to remove the fiber and debris. Let the solution sit still, which permits the flour to settle to the bottom. Pour off part of the

Cattails

water, then let it settle again. Repeat until a wet paste is left. Let this dry and store it until needed.

The roots may also have tiny buds growing out of them in various spots, used for branching out. Like the sprouts of spring, these can nipped off, peeled, and eaten either raw or boiled as a vegetable.

The "down," which is really minute seeds covered with a downy substance, can be stripped off the dry spikes. Upon separation they become quite fluffy in mass. A million or more "seeds" may be on each spike, and collectively they resemble a down. This fluffy material has been used for stuffing pillows and for upholstery. It has also been used for insulation, and during World War I it was used to make artificial silk and served as a substitute for cotton. It was also used by the American Indians for making absorbent pads; notably the Blackfoot tribe used it to dress wounds and burns. In India and Europe, the down is used for tinder. The leaves of cattail have also been used extensively to weave mats, baskets, and chair bottoms.

(See also BULRUSHES; PAPYRI; REEDS.)

cerimans A tropical fruit, the ceriman is aptly described by its scientific name, *Monsetra deliciosa* ("delicous monster"). A native of the American tropics, they are now raised commercially in Florida and California. Although sometimes found in other markets, the fruits don't taste good until fully ripe. At their prime, they compare well with mangos and bananas.

The ceriman is a vine of the Arum family, and it has large leaves that grow up to 3 feet wide. The fruit itself grows up to 12 inches long, and is more of a spadix than a stemmed fruit. It is covered with odd-looking hexagonal scales, which pop off when the fruit is ripe.

(See also TROPICAL FRUITS.)

chestnuts True chestnuts are members of the genus *Castanea* of family Fagaceae and have burrlike fruits that contain from one to three nuts. At one time, these nuts—gathered from the wild—were important as food in Europe as well as in North America. The nuts have also been significant as food for both wild and domestic animals. The wood of chestnut has been of importance to man as well, and the bark has been used in the manufacture of tannin. Here are a few kinds of chestnuts, along with some unrelated species that go by the name:

European chestnuts (*Castanea sativa*) Also called Spanish chestnuts, these trees grow in southern Europe, northern Africa, and east to the Caucasus, and they have been raised in India and Australia. The fruits produce three nuts inside the husk, and these vary in size. The largest ones are the more valuable commercially, and are called marrons. The nuts are eaten raw, or they are boiled or roasted. They are popular as stuffings for birds. Some of the smaller nuts are ground into a meal and, at one time, people in some parts of southern Europe more or less depended on chestnut meal for use in breadstuffs.

American chestnuts (*Castanea dentata*) Once these trees grew from Maine to North Florida, and west to the Mississippi. Unfortunately, a blight infected the trees at the turn of the 20th century, and they were all but wiped out. A few large trees remain, but these are scattered. Steps are being taken in some areas to establish the tree again in forests, but progress has been slow. At one time, the tree was important for food and as timber to early settlers as well as to the native Indians, and it was especially plentiful in the Appalachians and related foothills.

Allegheny chinkapins (*Castanea pumila*) These smaller cousins of the chestnut have only one nut in each fruit.

In addition to bearing a small fruit, the trees (or shrubs) aren't big enough to be valuable for timber. The nuts are, however, quite tasty, and are often eaten by foragers and rural folk. They are eaten raw or roasted, or sometimes used in other foods such as soups. After being roasted, the nuts are sometimes ground for use as a coffee substitute. The Allegheny chinkapin grows in the southeastern United States, usually in dry soil and sometimes in sandy areas along the coast. There are several subspecies.

Japanese chestnuts (*Castanea crenata*) These trees grow in the mountains of Japan, usually at altitudes of less than 3,000 feet. They are prolific, but bear rather small nuts. The nuts are usually eaten either boiled or roasted, and are common fare in Japan.

Chinese chestnuts (*Castanea mollissima*) This species grows in the mountains up to 8,000 feet. The nuts are good, and this plant is widely cultivated in the United States as well as in China. There are several varieties.

other chestnuts The Tahiti chestnut, *Inocarpus edulis*, grows in Fiji and other islands of the South Seas. As its scientific name indicates, it is not a true chestnut—but *edulis* means edible. The water chestnut or water caltrop (*Trapa bicornis*) of Eurasia and the water chestnut of South China (*Eleocharis dulcis*) are also edible, although they are tubers rather than nuts. *The U.S. Armed Forces Survival Manual* lists the water chestnut as a native of Asia that now grows wild in tropical and temperate regions of the world, including North America, Africa, and Australia.

Be warned that the European horse chestnut, *Aesulus hippocastanum*—a native of the Balkans—has a seed that looks like a chestnut, but it is highly poisonous. This plant, which has a beautiful flower, has been introduced to North America and other parts of the world as an ornamental. The Ohio buckeye, *Aesculus glabra*, of North America is also highly poisonous. The large seed was, however, highly valued by both the Indians and the white settlers as good-luck charms and for use in various cures. A buckeye in the pocket was believed to ward off rheumatism and hemorrhoids.

According to Michael A. Weiner's *Earth Medicine-Earth Foods*, a horse chestnut that grew in California was an important source of food to the Indians, in spite of the fact that one tribe ate it raw to commit suicide. Before eating the nuts, the Indians went through a rather complicated process of first roasting them for up to 10 hours in heated pits in the earth. They were then shelled, sliced, and placed in a freshwater stream for several days in order to leach out poison. Perhaps as a result of this procedure, the horse chestnut has also been used as a fish stupefier.

(See also NUTS.)

chickweeds (*Stellaria media*) These small flowering plants, native to Eurasia, now grow over most of North America and other parts of the world. Since they are widespread and tend to bloom throughout the year in temperate regions, they are an excellent food for survivalists to know. The young leaves are mild in flavor and can be eaten raw in salads or cooked as a potherb. The leaves are also scrambled with eggs or cooked in soups. Several other species of chickweeds are also edible.

chicories (*Cichorium intybus*) Native to Europe, chicories have long been eaten

Chicory

ular as a market vegetable, although they may be in the future. The plants grow wild over most of North America, and are considered weeds.

The roots are roasted and used as a coffee substitute or additive in both the United States and Europe. In England, most of the chicory is used with coffee. Commercial blends of ground coffee beans and chicory roots are widely marketed in the United States, and are especially popular in parts of Louisiana. Chicory adds a certain bitterness to coffee as well as a deeper color, and contains no caffeine.

The Belgian endive is a variety of *Cichorium intybus*. (Note: There is some disagreement over the scientific names of chicory and endive. Some sources hold that *Cichorium endivia* is correct for the endive.) Also, radicchio, or red chicory, is a cultivated variety of the same species. Popular in Italy, it is now available in some American markets. It is raised in Mexico and California as well as in Italy.

in the spring. Their young leaves, buds, and rootstocks are used raw in salads or the leaves are cooked as a potsherb. They have a bitter taste, which is especially pronounced in older plants. Often they are boiled in several changes of water to help remove the bitter taste; on the other hand, some people prefer them on the bitter side. The plants grow wild, and they are also widely cultivated in France, Belgium, Germany, and the Netherlands. In addition, they are raised during the winter in greenhouses and cellars, much like dandelions. In the 19th century, chicory plants were introduced to America, but they have never been pop-

chrysanthemums (family Compositoe) About 100 species of chrysantheums—and 3,000 garden varieties—grow around the world. The common chrysanthemum has been known for at least 2,500 years in China, where it probably originated and has been frequently used as a food or as an ingredient in various recipes. A modern book called *The Forgotten Art of Flowery Cookery* by Leona Woodring Smith, for example, lists a number of recipes that make use of chrysanthemum, including Peking turkey chowder, Hong King shrimp salad, dynasty eggs on toast, stir-fried chicken à la chrysanthemum, and Confucius sweet potatoes.

In addition to its use in food, the plant is important in various culture and folk beliefs. The ancient Chinese believed, for example, that a petal of a chrysanthemum flower, when put into a glass of wine, would help turn white hair black. The Japanese borrowed the chrysanthemum from China, and make even more use of it in cultural matters. The flower is important in the art and poetry of Japan. It is the national flower of Japan and, in A.D. 797, Mikado established the Order of the Chrysanthemum, a high honor.

In any case, the home gardener can now grow chrysanthemums all year-round as a potted plant. The petals are eaten raw in salads, or mixed in with chowders, casseroles, and other cooked dishes.

chufas (*Cyperus esculentus*) A nutgrass of the sedge family, the chufa has nutty tubers that are delicious. Chufas are eaten raw, or cooked. When dried, they can be ground into a flour, which is usually mixed in equal parts with wheat flour before making breads. The nuts can also be roasted, then ground into a coffee substitute.

Apparently, the chufa originated in Southern Europe, where it has been used for food for a long time. It is even cultivated. Chufa is also called rush nut, yellow nutgrass, earthnut, and ground almond. In the United States, it is sometimes planted for hog forage, and more and more it is being planted in wilderness areas for wildlife food. The American wild turkey is especially fond of chufa, which is usually planted in rather sandy soil, from which it is easier to be scratched up by the turkey and other two-legged foragers as well as by rooters. A number of other species of nutgrass grow around the world.

After the chufa was imported to America from Europe, the Indians started using it for food and in medicine. The Paiutes are said to have mixed the tubers of chufa (or some nutgrass) with tobacco leaves as a treatment for athlete's foot.

(See also ROOT AND TUBER VEGETABLES.)

cicelies (*Myrrhis odorata*) Native to central and southern Europe, these plants were at one time (especially during the 17th and 18th centuries) widely eaten as porherbs and were cultivated as such, often under the name sweet cicelies. Both the leaves and the roots are edible, but in modern times cicely has been dropped almost completely from vegetable markets and farmers' fields. It still grows wild, however, in pastures and along fencerows, often close to houses, from where it no doubt escaped from the gardens of yesterday. Waverley Root, who championed the plant somewhat, says that it is more assertive than chervil, and that in France it is called musky chervil. *Larousse Gastronomique* discussed "sweet cicely" as wild chervil, saying that it was more bitter than chervil.

Root also proposed that what is called cicely in America is another plant entirely and refers to herbs of genus *Osmorhiza*, especially to a much-neglected plant with a sweet-flavored root called sweet cicely. In *American Wild Flowers*, 2Harold N. Moldenke says, "Country boys and girls throughout almost all of North America are usually quite familiar with the thick, clustered, fleshy, aromatic roots of little woodland plants which they call variously sweetcicely, sweetjavril, sweetchervil, sweetanise, or aniseroot (*Osmorhiza*), and which they, like rabbits, like to nibble."

cilantro (*Coriandrum sativum*) The seeds of cilantro were one of the first spices used by man. They have been found in tombs of the pharaohs and even in ruins that date back to the Bronze Age. Today the seeds are used in curry powder, and are marketed as a spice under the name coriander.

In addition to the seeds, the leaves of cilantro are often used like parsley to flavor and garnish dishes. They are used in some countries to such an extent that they are more akin to vegetables and salad greens than to herbs. Cilantro is popular in modern Mexico and in the West Indies, but it is not widely used in the United States (although its use is increasing in some sophisticated circles and it has for a long time been cultivated by the Zuñi of the Southwest, who got it from the early Spanish explorers). Nor is cilantro widely used in England or France. Its strongholds include South America, Central America, China, North Africa, the Middle East, and Portugal, and it is native to the Mediterranean. It is sometimes called Chinese parsley or Mexican parsley. In Southeast Asia, it is a popular part of the cuisine, and the whole plant is used—leaves, stems, and roots.

citrus fruits The orange (*Citrus sinensis*) and most other forms of citrus fruits originated in the tropical and subtropical regions of Asia. Today citrus fruits of one sort or another are grown all over the world in areas of suitable climate. The orange, the lime, the lemon, and the tangerine (or similar easy-to-peel citrus such as the mandarin orange or the satsuma) need no introduction here, but perhaps it should be pointed out that the common grapefruit (*Citrus paradisi*) is a latecomer. It developed within the last 200 years, probably in Jamaica, and may have come from the pummelo by way of mutation. The pummelo (or shaddock) is the largest of the citrus fruits, and it can grow as big around as a basketball. It does well in lowlands around brackish waters. The ugli fruit was discovered in Jamaica and is apparently a cross between a sour orange and a tangerine.

A tangelo is a cross between a mandarin orange and a grapefruit. The sour orange, or Seville orange, is used mostly in cooking and for making marmalade. The citron (not to be confused with the citron melon) is raised primarily for its peel, which is candied. The pulp of some varieties of citron, however, is also eaten in Corsica and elsewhere. The true key lime doesn't have lime-colored skin; it is orange on the outside and lime on the inside. The kumquat, which grows not much bigger than a man's thumb, isn't really a citrus. It belongs to the *Fortunella* genus and grows in several varieties.

(See also BAELS; TROPICAL FRUITS.)

clover (family Leguminosae, genus *Trifolium*) Some 250 species of clover grow wild around the world; it is represented on every continent. The more agricultural species of the plant have been widely cultivated, partly because of its soil- improving abilities as a rotational crop. Today, the most widely cultivated species is the red clover *Trifolium pratense*, which grows from Siberia to South America. The clover is the state flower of Vermont, and, in the form of the shamrock, is the national emblem of Ireland. The plant usually has three leaflets on a stem; hence, the genus name *Trifolium*. St. Patrick is said to have used the clover leaf to demonstrate the unity of the Trinity to the Irish.

The peoples of Ireland and Scotland have made good use of the clover as food, and the American Indians were especially fond of it. The Indians even took the trouble to irrigate patches of wild clover so that they would produce well. The plant was so important that the Pomo Indians held a clover festival when the plants appeared in early spring. Reportedly, the ancient Chinese people also held a clover feast in spring, believing that a good mess of clover would cleanse the body's system.

The leaves of clover can be eaten fresh, preferably in a mixed salad. Be warned that too much clover consumed at one time tends to cause bloating. Some Indians cooked clover by piling it in layers and cooking it in a stone oven; others boiled the leaves with other greens. The blossoms are used, either fresh or dried, to make a tea, and the dried flowers together with the seeds can be ground and used in breads. The Irish make a bread by including a tablespoon of ground clover blossoms and seeds per loaf.

cocoa beans (*Theobroma cacao*) Both chocolate and cocoa powder come from the bean, or seed, of the cocoa tree. The beans are extremely bitter, however, and require much processing before being used. A native of tropical Middle America, the tree grows to a height of 40 feet and puts out football-shaped pods or fruits, which grow from 6 to 14 inches in length. Each pod contains from 25 to 50 beans, which are about an inch in diameter. The beans are about 50% vegetable fat, and of course cocoa butter, as it is called, is an important by-product.

The pre-Columbian Indians of lower Mexico, Central America, and tropical South America used the cocoa bean to make food and drink. In fact, the beans, which keep for a long time, were used as currency. Reportedly, 4 beans would buy a turkey (another American product), 100 a slave. The Spanish explorers found that the Aztecs revered a brown drink called *chocolatl*. Although the drink was bitter, the Aztecs thought that the beans came from the gods, and that drinking of *chocolatl* would bring wisdom. The Spanish took the beans to Europe, and in about 1550 sugar was added to the drink. Chocolate houses soon sprang up in England, Holland, and elsewhere. The English added milk in the 1700s, and chocolate candy, cake, ice cream, and other good foods have come forth ever since. Many other flavorings, such as anise and pepper, have been mixed with chocolate—and the Aztecs even used vanilla in it.

Today, raw cocoa beans are farmed in Africa as well as in tropical America. The manufacture of milk chocolate and other products is continued in Europe, North America, Australia, and India.

coconut palms (*Cocos nucifera*) Just as date palms are important to the wandering desert tribes of Arabia, coconut palms are important in the daily life on many tropical islands. These large palms do better when they grow near the sea, where they often lean out over the beach. Their large, meaty nuts float, and it is therefore not surprising that they are widely established throughout the tropics.

Of course, the coconut is now marketed all over the world, and coconut oil is a significant commercial product. (The oil is made from copra, or dried coconut meat; after the oil is extracted, copra meal is fed to livestock.) To many islanders, however, the uses of the coconut palm are truly staggering; in Indonesia it is said that the coconut

Coconut palm

palm has as many uses as there are days in the year. For example, coir, a fiber obtained from green coconut husks, is used in ropes, mats, baskets, and so on. The fiber is highly resistant to salt water, a highly desirable feature in making rafts for long sea voyages.

The sap of the tree is used to make a beverage called toddy. It is drunk fresh, fermented, or distilled. The mature nuts also yield a refreshing drink. The ripe nuts, if allowed to ferment, form a soft substance, called bread, that is also eaten. Of course, the bud of the coconut palm tree can be eaten as a vegetable, raw or cooked. The trunk of the tree is used for timbers, and the leaves are used for roofing as well as for baskets.

(See also DATE PALM; PALMS.)

corn salad (*Valerianella locusta* and other species) Also called mâche, lamb's quarters, lamb's tongue, and field salad, this plant grows wild in and around grain fields in Europe, North Africa, and Asia Minor. Also cultivated as a salad green, corn salad was imported to America, where it has become naturalized. The plant is not often eaten in America these days, but it is highly regarded in parts of Europe. It is cultivated in France, Italy, Holland, England, and other countries. In fact, it has been cultivated for food since Neolithic times, so long that its place of origin is not known for certain.

Wild corn salad has the reputation of being better than the cultivated form, possibly because the leaves don't keep well in supermarkets. In any case, the plant is usually used like lettuce, but it can also be cooked like spinach. It should be pointed out that the name corn salad came from the European term meaning grain, not from American maize.

cow parsnips (*Heracleum maximum*) These umbelliferous members of the carrot family are sometimes called cow cabbages and masterworts. Native to North America, they grow from Labrador to Alaska, south to California in the west and south to Georgia in the east. As the scientific name implies, they grow quite large (up to 10 feet high) and strong as compared to other members of the carrot family. In addition to eating the cow parsnip, the American Indians used it for many medicinal purposes, such as epilepsy and ear infections. The Indians also used ashes from the burned cow parsnip leaves as a salt substitute.

The cow parsnip is sometimes eaten by modern foragers and as a survival food. The young leaves can be eaten either raw or cooked. The stems can also be peeled and eaten, raw or cooked, and are said to taste somewhat like celery. The Indians also boiled the large roots for use as food, but some authorities hold that the roots are poisonous. Also be warned that cow parsnip resembles poison hemlock and water hemlock.

crab apples About 25 species of these flowering plants of the apple genus *Malus* (sometimes included in the pear genus *Pyrus*) grow in various parts of the world. Most of them have rather small fruits, usually no more than an inch in diameter, that are edible but too tart for most tastes. They do, however, usually make delicious jelly, pickles, and applesauces. In some wild species, the brightly colored fruits tend to hang on the tree throughout the fall and on into the winter, making them a valuable survival food. They grow wild in temperate regions of Asia and North America. The fruits can be sliced and dried for future use. Reportedly, the American Indians gathered the fruits in the fall of the year, put them into containers made of bark and buried them under the ground until spring, at which time they would have a sweeter taste.

Most of the planted crab apple trees are for ornamental purposes, and some have profuse blossoms. Only the Siberian crab apple (*Malus baccata*) is widely grown for its fruit. Also, the crab apple can be crossed with the standard apple, which is usually done to give an early fruit or one that can withstand colder weather. In Europe, a small, acidic apple is grown for making cider, and it is often called a crab apple.

(See also APPLES AND PEARS.)

crab grasses (*Digitaria sanguinalis*) People who try to eradicate this weed from their lawns will be surprised to learn that it was once cultivated not only in parts of Europe (to which it is native), but also in the United States. It was even called crop grass. The seeds can be ground into a flour, or eaten as a cereal or as a rice substitute. Even in modern times crab grass has been eaten regularly in parts of Africa.

Sometimes it is called finger grass because it has fingerlike stems atop the stalk. Several other species of grass also grow seeds that can be made into flour, or eaten in soups and stews either dried or fresh.

(See also GRAINS AND GRASSES.)

cranberries and blueberries   The American cranberry, *Vaccinium macrocarpon*, is cultivated on a large scale for its fruit, which has a long shelf life and is marketed fresh in supermarkets as well as being used in commercial cranberry sauce and juice. Cranberry sauce is something of a tradition in America, being served especially with turkey on Thanksgiving. It's a tradition that goes back to the Indians, who made and ate such a sauce with meats. The American cranberry also grows wild in the Northeast, west to Wisconsin. Moreover, several other species of cranberries grow wild, free for the picking, in North America, Europe, and northern Asia. These include the mountain cranberry, which is sometimes called lingenberry or foxberry. In addition to their use in cranberry sauce, pies, and jams, these wild fruits are sometimes eaten out of hand. The Indians dried them for use in pemmican.

Wild cranberries of one sort or another are sometimes marketed in northern Europe and elsewhere. The highbush cranberry is also used for fruit and sauce, but is not a true cranberry. Other edible wild species of *Vaccinium* have blue or 2blue-black fruits instead of cranberry red. These include the bog bilberry, the highbush blueberry (or swamp blueberry), and the late sweet blueberry.

Yankee Cranberry Sauce

Boil together one cup (236 milliliters) of maple syrup, one cup (236 milliliters) of apple cider, and two pounds (907 grams) of fresh cranberries. Reduce the heat and simmer the mixture until the cranberries pop open. Chill and serve with turkey or venison.

Some of the fruits are called huckleberries, but the edible huckleberry is really another genus (*Gaylussacia*). The confusion between them can be cleared up by taking a close look at the seeds. Genus *Gaylussacia* has 10 hard seeds, whereas genus *Vaccinium* has softer seeds and more of them.

The confusion in terms compounds on an international basis, since any number of fruits are called blueberries and cranberries. In South Africa, the "cranberry" and the "cape cranberry" are really a vine, *Aborbra tenuifolis*.

(See also HIGHBUSH CRANBERRIES.)

currants and gooseberries Some 120 species of the genus *Ribes* grow wild in North and South America, mostly in the Rocky Mountains and in the Andes. They also grow in Europe and the more temperate areas of Siberia, and have been widely cultivated in Scandinavia and other places that are too cold for raising grapes. During World War II, currants were raised in England because of a shortage of oranges and vitamin C. The juicy, piquant berries are usually used for making jams and jellies and pies, and they are sometimes eaten out of hand. The American Indians dried currants for use in pemmican and also ate the leaves,

which they gathered in early spring and boiled. Some modern practitioners eat the young leaves in salad or dry them for tea. America's early settlers made good use of wild currants, making wine as well as desserts with them. In Alaska, wild currant jelly or sauce is enjoyed with venison or other good meat.

At one time, the currant was a fairly popular fruit in America. But farming currants and gooseberries has been discouraged or even outlawed in some areas because the plants carry the white pine blister rust, a fungus that kills valuable timber trees. The red currant (*Ribes rubrum*), however, is one of the more common species and does not carry the rust as badly as the black currant. Consequently, the red currant is being raised more often in the United States for use in making jelly. This species is also marketed fresh, but is difficult to transport and store, which rules it out as a major supermarket fruit. The black currant (*Ribes nigrum*) is often raised in New Zealand, France, and England. As a rule, black currants are not often eaten fresh because they have a bitter taste, but they are popular in jellies and pies. In France, they are used in drinks and in *cassis*, a syrup flavored with black currant juice. The Scandinavians and the Russians also raise currants of one sort or another.

The gooseberry was once a very common fruit in America, used in various desserts, wine, and vinegar.

The gooseberry is still an important fruit in England and northern Europe, and is the main ingredient in recipes for gooseberry fool. Other gooseberry recipes are used, and gooseberry pie is said to have been one of Adolf Hitler's favorite desserts. Most of the hundreds of

British Gooseberry Fool

This recipe is distinguished by the use of real cream. Substitutions won't do.

1 pound (454 grams) gooseberries
1 1/2 ounces (43 grams) butter
4 tablespoons (57 grams) sugar
cream

Trim the tops and ends of the gooseberries and heat them in melted butter. Stir in 4 heaping tablespoons (57 grams) of sugar and heat until it is melted. Reduce heat, cover, and simmer until the gooseberries are soft. Measure this mixture, then measure an equal amount of cream. Whisk the cream until it is thick. Mix the gooseberries and the cream, and stir in a little more sugar if needed.

commercial European varieties came from *Ribes grossularia*, which once grew wild from North Africa to Scandinavia and east to the Caucasus. The so-called cape gooseberry that is raised in South Africa and elsewhere is really a ground cherry. And the "Chinese gooseberry" is really a kiwi fruit.

D

dandelions (*Taraxacum officinale*) Indigenous to Europe and central Asia, dandelions now grow wild in other parts of the world, including most of North America. They take their name from the French *dent-de-lion*, meaning lion's tooth, so named because of the toothlike shape of their leaves. The plants have been eaten for a long time, especially in Italy and France, and in America are still highly popular among foragers and enthusiasts of wild foods. Gathering dandelion—often from the lawn—is almost a springtime ritual for some people. The plants are surprisingly rich in vitamins and minerals, and are quite versatile.

The leaves are eaten raw in salads, or cooked as potherbs. They tend to have a bitter taste, which is appreciated by some people but not others. Cooking the leaves in more than one change of water helps get rid of the bitterness. The very young leaves tend to have a milder taste. The young roots can also be boiled and eaten, and old roots can be dried and roasted as a coffee substitute. The buds are also boiled and eaten, and even the mature flower can be cooked in fritters. The flowers can be used to brew dandelion wine, which is an old drink in England.

Some people grow the dandelion for food and, in order to have an early crop, they grow it in greenhouses or even basements. Early greens are cultivated and even sold commercially in Europe; according the the 1896 edition of Fannie Farmer's famous cookbook, *The Boston Cooking School Cook Book*, they have also been marketed in North America. In recent years, a new variety has been grown

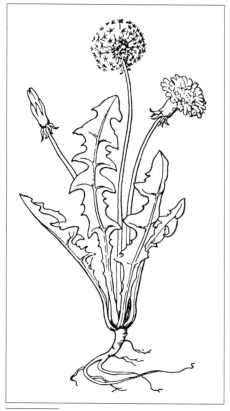

Dandelion

commercially. It has a longer leaf and is not as bitter as the wild dandelion.

A dandelion named *Taraxacum koksaghyz* was used as a commercial source of rubber in the former Soviet Union. Other dandelions have served as food for silkworms when mulberry leaves were not available.

date palms (*Phoenix dactylifera*) In the past, this species has been very important to the desert peoples across North Africa and Arabia. It grows wild between the Nile and the Euphrates, and is cultivated from the Canary Islands westward into India, as well as in America. Iraq, Algeria, and Tunisia are leading exporters of the fruit of the date palm. The tree grows to a height of 80 feet, and a single bunch can contain 20 pounds of dates. A good tree can produce 200 pounds per year. The date is high in sugar, and a similar tree in India provides fruit for the commercial production of date sugar.

The nomadic peoples of the desert made good use of the date palm, and not just of the fruit. Virtually the whole tree was used for timber, basket weaving, and fuel. The sap provided a beverage, drunk both fresh and fermented. The fruits were also used to make alcohol and vinegar. As with the other palms, the bud was used as cabbage. Even the seeds were ground and fed to camels and other livestock.

The palm tree was mentioned in the Bible, and the large leaves were used at times of rejoicing, as on Palm Sunday. Yet, the date itself was not mentioned. Perhaps its value as fruit was considered self-evident.

The value of the date palm to the desert tribes cannot be truly appreciated by modern man. Standing high, a single tree often served as a landmark, and a grove of palms usually grew in oases and at watering holes. Some 1,500 other kinds of palm trees have edible parts.

(See also COCONUT PALMS; PALMS.)

day lilies Native only to eastern and central Asia, day lilies have been cultivated for beauty and food for centuries in China. A cash crop even today, the flowers are dried and sold in Chinese markets around the world. It has been estimated that 4,000 pounds of the dried day lily blossoms are sold in New York City in a single year, usually under the name gum-jum or gum-soy.

Fresh blossoms are better than dried ones for most culinary purposes, but they don't stay fresh long enough for marketing. Home gardeners can have the best of both the aesthetic and culinary worlds. Since the blossoms drop off after a day or two, only to be replaced by others, picking a few for the table doesn't hurt a thing. The blossoms can be added to soups, casseroles, meat dishes, fritters, omelets, and so on. When raw the fresh blossoms, cut into ribbons, make a colorful and tasty addition to a green salad.

In the United States, the common day lily (*Hemerocallis fulva*) is a popular summer flower. It has escaped the flower bed and now grows along roads, in vacant lots, and along drainage ditches. Hence, the knowledgeable forager can enjoy day lilies from the wild as well as from the flower bed.

Further, the dried or fresh blossoms aren't the only parts of the plant that can be eaten. The young buds are good raw in green salads, or they can be boiled like beans. (But they should be cooked only for a few minutes and, for the best flavor, should be seasoned with a little butter.) The young buds are also delicious in stir-fry dishes. The inner young sprouts, found just as the plants are coming up, can be eaten raw in salads or cooked like asparagus. The underground tubers can also be dug up and eaten, raw or cooked, at any time of the year.

Although the complete forager will consider the day lily to be a plant for all seasons, it is not widely used for food except in Chinese cookery. But it should not be overlooked by anyone interested in organic gardening and natural foods. A hardy plant, the day lily isn't usually susceptible to diseases. It competes nicely with weeds, and normally doesn't require herbicides or insecticides.

Be warned that anyone who develops a taste for the day lily should remember to eat them in moderation. They can have a slight laxative effect.

docks Several species of dock (family Polygonaceae, genus *Rumex*) are eaten in America, Europe, and Asia. Most grow in temperate regions, but some extend to the fringes of the Arctic and, in fact, the Eskimo used to gather large quantities in summer and store them for winter use. The sour dock (*Rumex crispus*), which grows wild across the contiguous 48 United States, is a favorite plant among knowledgeable foragers. Other speces are also commonly eaten. The young leaves can be added to a salad, and often they are boiled. The older leaves may have a bitter or sour taste and should therefore be cooked in two changes of water, or perhaps mixed in with other greens of milder flavor.

The docks produce large numbers of tiny seeds on the top of the stalk. These can be ground into a flour and used in breads. The roots of dock are not normally eaten, but some species have been cultivated for the harvest of their root, which was used in the manufacture of rumex, a laxative drug.

dulses (*Rhodymenia palmata*) Red seaweeds, or algae, dulses attach themselves to rocks, large shellfish, or even other seaweeds. They grow in the Atlantic and Mediterranean. Their edible leaves, fan-shapped on a short root, grow to between 5 and 12 inches long.

In Ireland and Scotland, dulse is used in several ways. It is cooked fresh like a vegetable, or it can be dried. It is usually marketed in the dried form. In a traditional Irish recipe, dried dulse is soaked for 3 hours in cold water. Then it is simmered for 3 hours in a mixture of hot milk, butter, and pepper. In Ireland it is called dillisk or dillesk. In the Mediterranean, dulse is usually eaten in stews. Dried dulse can also be eaten raw, although it requires considerable chewing. It is, in fact, rather sweet when dried, and is sometimes used like chewing gum.

Irish moss, another red algae (*Chondrus crispus*) that grows on both sides of the Atlantic, is also edible. In shape, it resembles bib lettuce. Irish moss has been harvested commercially for use as carrageenan in pharmaceutical products. In fact, Irish moss is sometimes called carrageen.

durians (*Durio zibethinus*) These trees of family Bombacaceae which reach heights of 70 or 80 feet, grow in the Philippines, in the south of Thailand, and other parts of southeast Asia. They produce curious fruits, called ovoids, which grow as much as 8 inches in diameter and have hard external shells, which cover five more or less oval compartments. These compartments contain large edible seeds as well as a delicious sweet pulp. A sort of bonus, the seeds are removed from the fruit part and later roasted. They are said to have a chestnutlike flavor.

Although the durian is quite popular in its native regions, it has never caught on in with the rest of the world, no doubt because it literally stinks. The odor has been compared to that of rotten onions. According to Waverley Root, some Asian airlines will not allow passengers to bring durian fruits aboard, and some hotels will not allow them in guests' rooms.

E

edible valerians (*Valeriana edulus*) These plants grow in moist areas from Washington State to New Mexico, and around the Great Lakes. They are perennial herbs that stem from large roots. These roots can be dug and eaten in early spring, autumn, or winter. They have an unpleasant odor, but this disappears when they are steamed for a long time. The roots can be used in soups, or they can be dried and ground into flour for use in breadstuffs. The Indians cooked the roots for two days over hot coals in pits in the ground, which is not a bad method even today because of the long, slow cooking that is required for best results.

Several other species of *Valeriana* have been widely used in medicine both in the United States and in Europe. One common species, *Valeriana officinalis*, was sometimes called allheal. (It is also commonly known as garden heliotrope.) It has been used to treat epilepsy, nervous disorders, and other ills—and as bait for rattraps. It has been said that the Pied Piper carried the roots in his pockets to lure the rats out of Hamelin. In America, the Indians of the Northwest made use of valerian to treat wounds.

eggplants (*Solanum melongena*) Also called aubergines, eggfruits, and guinea squashes, these plants originated in southeastern Asia, where they have been cultivated for 4,000 years. The Arabs got the eggplant from the Far East, and the Moors carried it to Spain and, hence, to Europe. Thomas Jefferson is said to have raised eggplants on his farm in Virginia, and he may have been the first to cultivate it in North America. Today, the Italians of southern Italy and the Arabs of Turkey and other parts of the Middle East are the eggplant's greatest devotees. The peoples of the Middle East look upon the eggplant as "poor man's meat" and "poor man's caviar."

Strangely, throughout its history the eggplant has had a questionable reputation as table fare, and it was believed that anyone who partook of it would go mad or else suffer dire consequences, such as epilepsy. The great naturalist Linnaeus first named it *Solanum insanum* and later renamed it *Solanum melongena*. At one time or another, eggplants have been called "mad apples" or "apples of Sodom" or "Raging Apples." Even today, some experts recommend that eggplant be cooked well done.

egusis These African plants, sometimes called gourds and sometimes melons, are said to be a cross between pumpkins and gourds. In any case, they are not cultivated for use of their flesh, but rather for their seeds. The seeds, which are quite hard, are ground and used in soups and stews, or they are roasted and eaten whole. In African markets, they are available whole, canned, or powdered.

(See also GOURDS AND SQUASHES.)

elders Several species of elders grow in North America, Europe, western Asia, Siberia, and the Caucasus. These are shrubs or small trees that sometimes reach a height of 20 feet. All the elders have cream-colored flowers that grow in clusters, or umbels. These flowers turn into berries. The young limbs of the elder are filled with pith, and these limbs are often used for making toys, popguns,

---

Nigerian Snacks

West Africa's affection for snacking can also be seen in the streets of Nigeria, where people snack on egusi or melon seeds sold wrapped in bits of paper . . . .

2 tablespoons (30 milliliters) peanut oil
2 cups (454 grams) shelled egusi seeds
salt to taste

Rub the inside of a heavy iron skillet with peanut oil. Heat the oil until a drop of water sizzles when it hits the oil. Pour in the shelled egusi seeds. Cover the skillet and shake it until all the seeds have popped. The seeds should have a pale brown color. Remove them, salt lightly, and serve while hot.

—Jessica B. Harris, *Iron Pots and Wooden Spoons*

---

blowguns, whistles, and flutes. The wood of the larger trees polishes well and is sometimes used for making various objects, but in Denmark it was believed that the wood must not be used for furniture, and especially not for making cradles. This is because some people believe that Christ's cross was made from elder, and because others hold that Judas hanged himself from an elder. The American Indians used the elder, or parts of it, for various medicinal purposes. The Meskwakis, for example, used a tea made of the bark of the roots to treat headache and to encourage labor.

Two important American species—the common elder, *Sambucus canadenis*, and the blue elder, *Sambucus cerulea*—produce edible berries that are high in vitamin C. (The Mexican elder *Sambucus mexicana* and the black-berried elder *Sambucus melanocarpa* also produce edible berries.) The berries can be eaten raw but, more often than not, they are made into pies, jelly, wine, and fruit juice. Often they are mixed in with more acidic berries, such as sumac. The fully ripe elderberry can be eaten right off the bush, but it is sweeter if it is dried. The flower blossoms are also eaten, usually fried in fritters. They can also be used to make wine and a golden-colored tea.

Usually, the elder grows along streams and other moist places, and are often easy to spot along roadways. Elders are sometimes planted in gardens, both for flower and food. This practice probably derived from a European custom in which the elder was planted to keep evil away. It was considered to be bad luck to cut down such a tree.

Be warned that field guides to edible plants list the roots, bark, stems, leaves, and unripe (green) fruits as toxic, or at least questionable. The red berries of *Sambucus pubens* are also believed to be poisonous.

elkgrasses (*Xerophyllum tenax*) Although not listed in some guides to edible American plants, this species grows on the ridges of the Northwest from California to British Columbia. It is also called beargrass, bearlily, pinelily, firelily, and squawgrass. It makes a large bunch of grass, out of which grows a tall stem. The last foot or so of the stem is covered with thousands of white flowers. The Indians of the region used fibers obtained from the leaves of elkgrass to weave clothes and baskets. They also roasted the large rootstocks for food.

(See also GRAINS AND GRASSES.)

evening primroses (*Oenothera biennis*) These plants and other related species in the large genus have edible taproots and leaves. The roots, which grow up to 2 inches in diameter, can be peeled and boiled. Usually only the first-year roots are eaten, before the stalk goes to flower. Nonetheless, these roots can be gathered in autumn and winter, which makes them a good survival food. The roots are also quite good, having a sweet taste. The leaves and young shoots are boiled as potherbs and are also eaten raw, usually in salads. The evening primrose is native to North America and, because of its early popularity as a forage food, it was introduced to Europe in 1614, where it was cultivated for food in Germany and Spain. In Poland the roots were used to fatten hogs. The plant now grows wild in most parts of Europe. In early America, the roots were boiled in syrup until candied, and they were also boiled with meats.

At dusk the plant's buds quickly unfold into a large flower. A Dutchman named Hugh de Vries studied a variety of the evening primrose (*Oenothera lamarckiana*) in his garden. His observations led to an important theory of mutations, which meant that evolutionary changes could take place quite suddenly and did not necessarily have to be made by small changes over geologic ages.

F

feijoas (*Feijoa sellowiana*) These small fruit trees of the myrtle family are native to the mild, rather arid regions of South America, south of the Amazon Basin. Often called pineapple guavas, feijoas are cultivated in New Zealand and Australia. Although the damp climate of Florida is unsuitable for their culture, they were raised on a limited basis in parts of California in the early 1900s, and in southern Europe as early as 1890.

The 2-inch oblong fruits fall when ripe, and must be kept in a cool place until they are soft enough to eat. Their taste is tart and good, rather like pineapple mixed with muscadines. Feijoas are eaten out of hand as well as being used in various fruit salads and similar dishes. They are also used for making pies and jellies, and are crystallized. Feijoas have large white flowers, reddish purple inside, that are also edible.

fenugreek (*Trigonella Foenumgraecum*) Virtually unknown in America, fenugreek has an exotic flavor that is valued in some cultures. Indigenous to the Mediterranean regions, it has for a long time been cultivated in parts of Europe, western Asia, India, and North Africa. This legume has slender, curved pods that contain tiny, hard seeds. When ground, these seeds are used in curry powder and chutneys as well as flavoring agents.

In Ethiopia, the seeds are cooked as vegetables. In India, the leaves and young pods of fenugreek are used as potherbs and, of course, they are used in curries in India, Pakistan, and other

places. In Israel, the seeds are used to flavor sauces. In Egypt and other parts of North Africa, the seeds are ground and used in making breadstuffs, sometimes being added to regular wheat flour or cornmeal. Being high in calories, protein, and calcium, the seeds are believed to help people put on weight and to help young mothers produce more milk. They are also believed to have medicinal value, and are used for poultices and emollients. In addition, the plants are used as hay to feed livestock.

(See also HERBS AND SPICES.)

fiddleheads A large number of ferns grow around the world and the new fronds of most species look somewhat like the head of a fiddle. Some survival manuals say that these fiddleheads are edible, noting that the common bracken grows throughout the world except in the arctic regions. Other authorities warn that the common bracken and some other ferns may be carcinogenic. According to *Uncommon Fruits & Vegetables*, the fern curator of a large botanical garden has stated that only the fiddlehead that is commonly marketed is definitely noncarcinogenic. Specifically, this is the ostrich fern, *Matteuccia struthiopteris*, which grows from Virginia to Newfoundland. It is available in the wild, and is also marketed in fresh, canned, and frozen form. If they are gathered at the proper time, fiddleheads taste somewhat like asparagus and artichokes. The Japanese are especially fond of fiddleheads of one type or another.

In spite of the modern warnings, various species of ferns have been eaten for pleasure and for survival in many parts of the the world, especially in the

190

tropics. The common bracken fern (*Pteridium aquilinum*) is eaten, both cooked and raw, across the Northern Hemisphere in Europe, Asia, and North America. It is also eaten in New Zealand and Japan. Usually, only the young fronds of bracken are eaten because the older ones (longer than 8 inches) have a bad taste and may be more or less toxic.

In addition to having eaten the fiddlehead fronds of the bracken fern, some of the American Indians also made bread from the ground roots. The Hopi Indians ate the rootstocks as well as the stems of the horsetail fern. According to Michael A. Weiner's *Earth Medicine-Earth Foods*, the Indians of Washington baked the peeled roots of the sword fern with salmon eggs in pits in the ground.

figs One of the earliest of cultivated fruits, the common fig, *Ficus carica*, originally grew from Turkey to India. It spread to the Western world, and Greece is said to have obtained the tree from Caria (thus the second part of its scientific name). The Romans made especially good and wide use of the fig as table fare. It was eaten not only by epicures but also by slaves and common folk. The fig is also mentioned a number of times in the Bible, both figuratively and literally, and of course Adam and Eve are said to have covered their nakedness with fig leaves.

Today the fig is widely cultivated and marketed in fresh and dried form, as well as in such items as fig preserves and Fig Newtons. The fig is especially plentiful in the Mediterranean area. In Greece, the figs are flattened and strung on reeds before being taken to market; hence, the name string figs. In addition to *Ficus carica*, a number of other species or varieties are raised in one region or other. In India, *Ficus benghalensis* has heart-shaped leaves that are used for elephant food, although the fruits are not outstanding and are eaten mostly in times of scarcity. Also, India's *Ficus elastica* doesn't yield edible fruit, but it is used for rubber.

In the wild, about 800 species of fig trees grow in tropical and subtropical lands. Some have woody fruits that are inedible. But most of the figs are soft and make excellent survival food or fruit for the forager.

(See also TROPICAL FRUITS.)

fireweeds These plants (*Epilobium angustifolium* and related species) often spring up in astounding numbers after a fire burns off woodlands and fields. Hence, their popular name. One species or another grows in southern Canada and the northern part of the United States, as well as in parts of Europe and Asia.

Fireweed is fairly popular among foragers partly because it is easy to identify (in areas blacked by fire, the red blossoms can be spotted from some distance away) and because it grows thickly in burned areas, so that a mess can be gathered quickly. In spring, the young shoots are eaten like asparagus. The young leaves and buds are cooked as potherbs, or eaten raw in green salads. Dried leaves can be used for making tea. When peeled, the old stalks reveal a pith that can be chewed or used in soups. The pith can also be used in making breads.

Fireweed gets mixed ratings in the literature on edible plants. The *Field Guide to North American Edible Wild*

Plants reports that it is not a very tasty food. *Earth Medicine-Earth Foods*, on the other hand, says that the plants make excellent vegetables, and that the French Canadians value the fireweed, which they call *asperge* or wild asparagus.

fruits See CITRUS FRUITS; STONE FRUITS; TROPICAL FRUITS.

fungi See JEW'S EARS; MUSHROOMS; SMUTS; TRUFFLES.

G

galls Most people will be surprised to learn that galls, the very common wartlike or gouty growths on plants and trees, are sometimes eaten and can in fact be highly nutritious. Not long ago in America, the settlers and farmers in Missouri and Arkansas fed hogs and other livestock with a small gall that grows on a scrub oak of the Ozarks. Sometimes a single tree will contain thousands of galls, so that gathering them is easier than might be expected.

Even humans eat galls, some of which are quite palatable. In Mexico, a large growth caused by a gall wasp is gathered and sold in the fruit stands of Mexico City. In Crete, a sweet, aromatic gall that grows on a sage plant is marketed. In the United States, the catmint gall has been eaten in the past, and it takes on the pleasant taste and aroma of the catmint plant.

There are over 2,000 different kinds of galls in the world. These are made by wasps, ants, moths, beetles, amphids, flies, and mites. In all cases, the insect injects a substance into the plant in one way or another, which causes an abnormal growth. Usually, this growth is used by the insect to house its eggs and larvae. The gall provides food as well as shelter, and some of the growths produce large amounts of sugar. Some even exude a honeydew, which is gathered by bees and the honey ant.

Some galls contain lots of tannin (up to 65%), and have been harvested commercially all over the world for use in the manufacture of inks and dyes. From the time of the Greeks, the Aleppo oak of Asia Minor provided the galls that were used in making ink for important documents. Even in modern times, this ink was specified for use in the printing done for the United States Treasury and the Bank of England.

garlic See ONIONS AND GARLIC.

garlic mustards (*Alliaria officinalis*) Indigenous to Europe, North Africa, and temperate Asia, these edible plants now grow wild in America from Canada down to Virginia, west to the Dakotas. The taproot, when crushed, smells somewhat like garlic, but it's the tops that interest enthusiasts of wild foods. In spring or early summer, the young leaves, blossoms, and seedpods can be cooked or eaten raw in salads. These parts can also be used in soups and meat casseroles. In Europe, *Alliaria officinalis* is also called hedge garlic, Jack-by-the-hedge, and sauce-alone.

(See also BLACK MUSTARD.)

ginkgo nuts The fruits or nuts of the ginkgo tree (*Ginkgo biloba*) have a gooey outer layer that has a bad smell and irritates the skin. This layer covers a hard white shell, which in turn covers a nut. The nut itself is covered by a brown inner skin, which is easily removed after being plunged into hot water. What's left is edible. Although the ornamental ginkgo tree grows in Asia, North America, and Europe, the nuts are not commonly eaten, perhaps because of the time required for preparation. In China, the plant has been grown around temples since ancient times and is considered sacred. It is also called a "living fossil," since it has existed unchanged for millions of years. (See also NUTS.)

ginsengs Both the American ginseng (*Panax quinquefolius*) and the Asian ginseng (*Panax ginseng*) are in demand even today in oriental markets, not as a food, but as an important ingredient in medicines and cure-alls. In China especially, ginseng has been been valued since recorded history, and is supposed be a heart stimulant and an aphrodisiac. The plant has long parsniplike roots, which sometimes look like a man. (The name ginseng in Chinese means "like a man" and the generic name panax comes from the Greek, meaning panacea or "cure-all.") On the market, the value of the root depends largely on its shape. Those that look the most like a man are of more market value; a few are almost priceless. Since the supply of wild ginseng has been greatly diminished, the plant is now cultivated on a limited basis. The wild plant, however, brings a better price on the market. At one time, ginseng was an important export for North America, where it grew in rich woods from Quebec to Florida. During the 1700s, gathering wild ginseng was a profitable occupation, which the Indians engaged in. In 1773, the sloop *Hingham* set sail from Boston to the Orient with 55 tons of ginseng.

These days, the main market is still in China, but in the past the plant has had great commercial value in England and France. The American Indians also valued the roots of ginseng, which they used in remedies for nausea and other ills. The Meskwakis made a love potion by mixing root of ginseng, snake, mica, and gelatin. The ginseng plant is not widely eaten, possibly because of its market value, but the roots can be boiled and eaten as a survival food. The young leaves (either fresh or dried) can also be used to make a tea.

glassworts These members of the goosefoot family comprise a dozen species of herbs that live in or near the salt marshes of Europe, Asia, and North America. In some areas, the plants grow in rocky places and cliffs along the sea. One species, *Salicornia herbacea*, is high in sodium content and was at one time widely used the manufacture of soap and glass.

Glasswort is also called samphire as well as sea bean, Peter's cress, sea fennel, beach asparagus, pickle plant, marsh samphire, sea pickle, and pousse-pied. It is also called saltwart because it grows best near salt water, including the inland salt marshes of the American West. In fact, the Indians of Utah and Nevada used glasswort seeds to make flour.

By whatever name, glasswort has been widely eaten in Europe as well as in America. Usually, the stems are boiled as a vegetable or pickled in vinegar. (Although the plant grows wild up and down American coastlines, the pickles are imported from France.) The young shoots and tender leaves can be eaten raw, preferably in a mixed salad, to which they add a pleasingly salty taste. The plants are best in spring, but they can be eaten all summer and into autumn. Because they are easily recognized and are edible for a long period of time, glassworts are a good plant for the survivalist to know. The leafless stems are joined like cactus and look like bird's feet.

(See also HERBS AND SPICES.)

goa beans (*Psophocarpus tetragonolobus*) The unusual goa bean plant is grown in the tropics and has recently been the subject of international study as a food. It's a climbing plant, covering shrubs and trees, that produces a 9-inch-long bean. Often called the winged bean, it has a square shape with fins sticking out from the corners. Both the green bean and the seeds are edible. The leaves are also edible, cooked or raw. The seed can also be used to make a cooking oil. Moreover, the plant has root tubers and shoots that are edible. All of these parts are good sources of protein.

The goa bean plant is resistant to disease and has a high yield, both important considerations in large-scale farming. However, the plant doesn't keep well, and it doesn't grow too successfully outside the tropics. Hybrids are being developed, though, and in recent years the plant has been the subject of international study as a food source. Although this bean is not yet widely available in the supermarkets of the United States, it is being grown in Florida, California, and Texas. It could well become important as hybrids are developed to withstand cool climates.

In addition to the goa bean and winged bean, other names include asparagus bean, four-angled bean, manila bean, and princess pea.

good King Henries (*Chenopodium bonus-henricus*) These perennial herbs of Europe have also been naturalized in America. They grow to about 2 feet in height and have leaves shaped like arrowheads. These leaves are cooked and eaten like spinach; usually two changes of water are needed to take out some of the bitter taste. The young shoots are also eaten like asparagus or early cattail. The plants have been cultivated for food,

sometimes under the name mercury or all-good.

(See also HERBS AND SPICES.)

gooseberries See CURRANTS AND GOOSE-BERRIES.

gourds and squashes The gourd family, Cucurbitaceae, contains about 700 species of climbing herbs, including the watermelon, the cucumber, and the squash, as well as gourds. There is some confusion about what's a squash or a pumpkin and what's a gourd, but in general the former is native to the New World, and for the most part the latter is native to the Old World. As a vegetable, squashes were cultivated for at least 2,000 years in America before the white man arrived. These vegetables were very important, right along with beans and maize (corn). In the Old World, various melons were enjoyed and the gourd was eaten on a limited basis. The various summer squashes, winter squashes, pumpkins, and so on are now cultivated in Europe and elsewhere, but have never quite reached the level of importance that they had to the American Indians. The French never took too kindly to squash and the British tended to called it vegetable marrow.

Primitive man also used the gourds and hard-shelled squashes as dippers or laddles, utensils, floats for fishnets, and musical rattles. Gourds have been found in Egyptian tombs, and in middens in Peru, both dating back to 3000 B.C. Although exact classification is sometimes difficult or confusing, in general the edible American squashes are *Cucurbita pepo* or its cousins, and the gourds are *Lagenaria siceraria* or its cousins. In any case, here are some interesting kinds of gourds and squashes that are eaten today:

calabazas   Sometimes called West Indian pumpkins, these large squashes are usually marketed by the slice or wedge. They have beautiful yellow flesh and a delicate flavor. Their scientific name is *Cucurbita moschata*, but there are several varieties, variously called Cuban squash, toadback, zapallo, abóbora, and giraumon. A wild variety once grew in Florida, and it was widely used by the Indians and the early settlers. Today, calabaza is raised commercially and marketed in Central and South America and, of course, in the West Indies.

Calabaza can be cooked in a number of ways, but it is usually put into soups and other dishes or puréed and used in pies and other recipes that call for pumpkin. The seeds are first boiled for a few minutes and then roasted.

spaghetti squashes   A strange variety of the regular squashes (*Cucurbita pepo*), these squashes contain many strands of fibers. When cut, they look like melons stuffed with noodles. They are cooked and eaten in various ways, which include a desert recipe. The spaghetti squash, or spaghetti vegetable, is now raised commercially in Mexico and the United States. There is also a similar edible vegetable that is sometimes called a spaghetti gourd.

chayotes (*Sechium edule*)   Also called mirlitons, vegetable pears, custard marrow, xuxus, and several other names, these members of the Cucurbit family of vines grow an edible squash the shape and size of a pear. The squash has a crisp fine flesh, and can be cooked in a number of ways. The large seed in its center is also edible. A native of Mexico, the chayote is also eaten in other parts of America. Popular in parts of Florida and Louisiana (where it is usually called mir-

liton), it is increasingly available in supermarkets in other parts of the country.

sweet dumpling squashes   These are another cultivated variety of the ordinary *Cucurbita pepo*. About the size of small cantaloupes, the squashes have a yellowish flesh with a slight sweet flavor. According to Elizabeth Schneider's *Uncommon Fruits & Vegetables*, the variety was developed in Japan by a seed company, who called it a vegetable gourd in its catalog.

squash blossoms   Knowledgeable home gardeners have for a long time enjoyed the blossoms of the various squash plants. They are often dipped in a light batter and fried, or sautéed and used in omelets or other recipes. (Note that the male blossom doesn't actually produce a squash or grow on the end of an immature squash, so they are the best ones to pick for the table.) Squash blossoms are sometimes available in modern markets, but they have a short shelf life. The American Indians, who made extensive use of the blooms of squash and pumpkins in soups, also dried them for future use.

edible gourds   Although most Americans tend to think of gourds as being ornamental instead of as food, this is not universally the case. Several gourds are eaten in various parts of the world. According to Waverely Root's *Food*, the fuzzy gourd (*mao gwa*) is preferred in China to squashes, and in Africa and Madagascar, the bottle gourd (*Lagenaria sicetraria*) is eaten—fruit, leaves, and shoots. Also in Africa, the gourd called egusi is raised for its edible seeds. *The U.S. Armed Forces Survival Manual* lists the edible "wild desert gourd" as being abundant in the Sahara, Arabia, and parts of India, and says that it also has edible

---

Gathering Wild Grass Seeds

Collecting the seeds of wild grasses was undoubtedly a widespread practice, providing a major source of food to many of the world's peoples before the rise of agriculture. Indeed, it is widely believed that collecting wild grass seeds led to the beginning of cereal cultivation in West Asia, the Mediterranean, the Sudan, and Ethiopia, and probably China; and the same is true of the origin of maize cultivation in Mexico and Central America. That agriculture began where wild grasses grow is one of the principal reasons why so few of the hunting and gathering peoples still live where wild grasses are abundant enough to be a major factor in their food supply. . . .

The technique of harvesting wild grains was very simple, and specifically women's work. They did not reap the grasses, but simply walked through the stand, each woman holding a basket in one hand and a stick in the other. She would place a handful of stalks against the edge of her basket, and beat the seeds into it with the stick.

—Carleton S. Coon, *The Hunting Peoples*

---

flowers and seeds that can be roasted or boiled. The same manual also lists the "wild gourd" or "luffa sponge" of the tropics as survival fare, since it has edible fruits (when immature), shoots, flowers, leaves, and seeds. This strange gourd (*Luffa cylindrica*) contains a fibrous interior which, when dried out, is used as a sponge. It is also called dishcloth gourd and rag gourd.

squirting cucumbers (*Ecballium elatarium*) These odd gourds, which eject their seeds with a squirt of fluid, are edible. They grow in the Mediterranean regions.

(See also EGUSIS.)

grains and grasses Most of the world's breads and cereals come from cultivated grasses, or at least plants that have evolved from wild grasses. The long list includes wheat, oaks, millet, and even maize (which is called corn in America). Wild rice and other grasses were used extensively by the American Indians, and the Spanish word *piñole* applied to any

meal or flour made from grass seeds. Often, the seeds were parched before being made into flour. And, as often as not, a number of different kinds were mixed together.

Some grains such as wheat, maize, and rice are used in most parts of the world, but other grains remain rather regional. Naturally, some grains grow better in certain climates than in others, which is why teff is an important grain in parts of Africa but not in such wheat-producing areas as Kansas. Although more and more kinds of grain are being made available in modern gourmet or health-food markets, the trend in recent history has been to specialize in grains that can be grown, harvested, and processed on a large scale. As a rule, the more popular grains are available at such low prices that modern man has pretty much forgotten about the other grains and grasses. Few American farmers, for example, realize that common barnyard grass, crab grass, and "weeds" such as green amaranth produce edible seeds that can be made into bread. The trend is toward more and more use of machines and chemicals and less and less use of manpower in reaping, sowing, and processing grain. The price of mass production is the increasing use of, and dependency on, commercial fertilizers, pesticides, and herbicides.

In addition to being used in breads and cereals, many grains have become important in the production of beer and, of course, as food for livestock, which in turn are eaten by man.

(See also CRAB GRASSES; ELK GRASSES; TEFFS; WILD RICE.)

grapes (genus *Vitis*) Some 60 species of grapes are native to the north temperate zone, and many of them grow wild in North America. Some of these, such as muscadine, make excellent eating and others are more often used for jellies, pies, and beverages. The leaves of most species are also edible, especially if they are gathered while tender in early spring. The vines can sometimes provide safe water for drinking in emergencies. Be warned that the poisonous moonseed also bears a cluster of fruit that looks like small grapes. As the name implies, this fruit has moon-shaped seeds whereas grape seeds are roundish.

Grapes have been an important fruit for mankind, and, of course, their juice has been widely used for making wine. Grapes have been cultivated for food since recorded history, and seeds of *Vitis vinifera* have been found in the remains of the Swiss lake dwellings, dating back to the Bronze Age, as well as pre-Hellenic Greek ruins and in mummies of ancient Egypt. Noah is said to have planted a vineyard, and grapes were cultivated in the hanging gardens of Babylon. It is guessed that modern viticulture had its beginnings somewhere around the Caspian Sea.

Besides providing fresh fruit, grapes are now used to make raisins as well as wine and grape juice. The first raisins are said to have been made in Asia Minor by burying grapes in the sand. In addition to preserving the fruit, making them into raisins also concentrated the sugar. In the past, the grape was in fact an important source of sugar, and its juice was used in concentrated form. In Europe the seeds of grapes have also been used to produce cooking oil.

(See also MUSCADINE.)

grasses See GRAINS AND GRASSES.

greenbrier (*Smilax bona-nox*) Sometimes called bullbrier, these climbing vines of the southeastern part of the United States grow in woods and fields. The vines grow from large tuberous roots, and often reach high into the trees. The mature vines have sharp stickers. They also have tendrils, young leaves, and berries. The young shoots, or the very tips of the vine or its branches, are often eaten raw as a trail nibble or in a salad. They are also cooked as a vegetable or used in casseroles. The very young leaves can also be cooked like spinach, along with the young shoots and tips. The tendrils can also be cooked. In taste and texture, greenbrier makes one of the best wild vegetables and is a good addition to a tossed salad.

The tubers contain starch. When sliced, pounded, and washed, a sediment will settle to the bottom. This can be dried like flour and used in breadstuffs, or as a thickener for stews and soups. It is also used to make a drink (hot or cold) and even jelly.

Stir-Fried Greenbrier

While hiking in the woods or along the edge of fields, pinch off the young tips of the greenbrier vines or similar *Smilax* vines. These may vary from matchstick size to tips as large as a man's little finger. The key is to get only the fresh, tender tips.

    greenbrier tips
    water
    salt
    butter or margarine

Boil the tips for 3 minutes in slightly salted water. Then melt a little butter in a skillet or wok and sauté the tips for a few minutes. Salt to taste.

Several other species of *smilax* vines are also eaten. The carrion flower (*Smilax herbacea*), which grows in the northeastern United States and south to northern Alabama and Georgia, and west to Iowa, is similar to greenbrier. (See also ROOT AND TUBER VEGETABLES.)

greens This popular culinary term applies loosely to leaf vegetables, such as spinach, that are cooked as a potherb or, sometimes, put into salads. (The term salad greens is not uncommon, either.) People who forage for wild foods as a hobby are fond of making a green salad from dandelion leaves, watercress, chicory, and many other wild plants. Also, such greens as pokeweed leaves, dock, and wild mustard are often mixed together and cooked in a pot. As a rule, the young tender leaves are "better" than the older ones, which tend to be bitter and are usually cooked in more than one change of water. On the other hand, some people like the bitter taste.

The tops of many of the common root vegetables are also enjoyed as greens. Turnips, for example, are in some parts of the American South valued more as "greens" than as roots. Beets and radishes also have edible tops, and Dr.

George Washington Carver, the black scientist who helped make the peanut an important crop in the United States, also did lots of work with the sweet potato, pointing out that the green leaves are very good, plentiful, easy to grow, and nutritious. But, unfortunately, not many people eat sweet potato leaves, and this is a good example of a food resource that goes to waste.

Believe it or not, the Eskimo and some other primitive peoples made good use of the greens that had been eaten by caribou and other cud-chewing animals that were killed for meat. The contents of the stomach are said to make excellent greens, having a slightly tart flavor, as if sprinkled with vinegar. (See also BLACK MUSTARDS; PURSLANES; WATERCRESS; WATER LETTUCES.)

ground cherries These unusual plants are kin to tomatoes, not to cherries. In fact, the fruits of some species are raised commercially and sold as husk tomatoes. In Mexico, the *tomatillo*, or Mexican green tomato (*Physalis ixocarpa*), is an important part of the cuisine and is used in salads, salsas, and other dishes, cooked and raw. It is especially good in stir-fry dishes. The *tomatillo* was cultivated by

---

The Complete Forager

In their respective seasons Professor Carver had for a long time included wild vegetables in his own diet—clover tops, dandelion, wild lettuce, chicory, rabbit tobacco, alfalfa tops, thistles, bed straw, pepper grass, wild geranium, purslane, hawkweed, Flora's paintbrush, water cress, shepherd's pruse. The dainty chickweed, which was pretty as to appearance and delicate as to taste, could be happily combined in salads or stews, cold or hot. Curled dock made as good a pie as its cultivated cousin the cultivated rhubarb. From sour grass, or old-fashioned sheep sorrel, he made not merely pies but confectionery and paint.

A little earlier than asparagus in most localities, the tender shoots of the pokeweed poked up through the soil. These were as delicate and delicious as the asparagus tips, which they resembled, as was also the swamp milkweed. Even primrose, so pink as to be almost red, grew in masses.

Lamb's quarters, choicest of vegetables, was scattered over the temperate and subtropical sections of the country, and was available from early until late summer. It made an immense amount of green stuff, was tender, crisp, and cooked easily. Before McCollum tested it on his rats and announced it contained the Fat Soluble A—one of the newly discovered and much-needed properties called "vitamins"—Professor Carver had proclaimed its high medicinal value.

—*George Washington Carver* by Rackham Holt

---

the Aztecs, and in recent years it is becoming more widely available in the United States. Another species is called the Cape gooseberry (*Physalis peruviana*). A native of Peru, it is now cultivated in South Africa, India, Australia, China, and New Zealand. Available in markets, the Cape gooseberry is eaten raw or cooked.

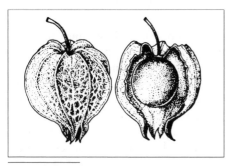

Ground cherry

Similar fruits grow wild in many parts of the world, and a species (*Physalis pubescens*) grows in the eastern part of the United States, west to the Rocky Mountains, down through Mexico, and on into South America. Fruits of similar plants are called *poha* in Hawaii, where they are widely used in jelly and jam.

The wild fruits are usually rather small, about the size of marbles. They are easily identified by the paper husk that grows over the fruit, which closely resembles a miniature Chinese lantern. The fruits can be eaten raw, either out of hand or in salads. They can be stir-fried with other vegetables, or used in pies, jelly, jam, or relishes. Ground cherries can also be dried for future use.

Actually, several species of *Physalis* grow in the wild, and the fruits may be yellow, red, purple, or blue-black when ripe. Regardless of color, all of the ground cherries can be identifed by the lanternlike husk; but some forager's guides recommend that only ripe fruits be eaten.

groundnuts (*Apios americana*) Also called Indian potatoes, these vines were critical to the American Indians and to the Pilgrims, who more or less lived on their tubers during their first few winters in New England. The egg-shaped tubers, growing up to 2 inches long, string out along the root and are easy to dig. (A similar species, *Apios priceana*, has only one tuber.) The tubers are eaten either raw or cooked, and are excellent when sliced and fried. The vine also produces a beanlike pod, which has seeds that can be eaten like peas. But the tuber is the important part. Able to be dug and eaten all the year, it is an excellent survival food within its range over the eastern part of North America, from Canada to Texas, west almost to the Rocky Mountains.

If the Irish potato (from South America) had not been transplanted to Europe and then back to North America, the groundnut might well have become an important commercial crop. In fact, it was cultivated in France for a time. Today, it is usually eaten only by foragers of wild foods.

(See also ROOT AND TUBER VEGETABLES.)

H

hackberries (*Celtis occidentalis*) These popular shade trees of the elm family are easy to recognize because of the wartlike growths on their trunk and, usually, by the many small galls on their stems. They bear in profusion a small berry that is edible and that contains sugar. (Hackberries are often called sugarberries.) While the purplish black berries have large seeds and are not widely eaten these days, they can be valuable as a survival food because they tend to stay on the trees during the winter—unless robins and other birds get them all. The American Indians made good use of hackberries, and even ground dried seeds for use as a seasoning.

Actually, several closely related species are all edible, and the quality seems to depend partly on the weather. The berries can be eaten out of hand, fresh or dried, and can also be used to make pies and jellies.

The common hackberry grows in the eastern part of North America, and it is also planted in Europe and elsewhere as a shade tree or decorative tree. Several other species of hackberries grow in the tropical zones and in some temperate regions of the world. The trees have been used for furniture and other purposes, sometimes under the name beaverwood. Because of the large number of berries, hackberries are a valuable source of food for nonmigratory game birds such as the bobwhite quail and the grouse.

haws See HAWTHORNS AND HAWS.

hawthorns and haws These members of the rose family (Rosaceae), genus *Crataegus*, grow in Europe, North Africa, and Asia Minor, but the largest number are native to North America and many of them have edible fruits. Usually, these trees are small with dense branches that have sharp spines. The fruits generally ripen in autumn, but this rule does not always hold. The southern mayhaw (*Crataegus aestivalis*) ripens in late spring, is popular in east Texas (and elsewhere) for making jelly, and is the subject of the annual May Haw Festival in Colquitt, Georgia. In any case, the berries of the various species can be eaten out of hand, with the ripest ones being the sweetest. The berries are also crushed and brewed in a tea, and, in general, are used like rose hips. But the main use of the various haws are for making jelly. In addition to being used fresh, they can be dried for winter use, and the American Indians used them in pemmican.

In England, the hawthorn has been used in folk medicine, and some people attach religious significance to the tree because it was believed to have been used to make the crown of thorns worn by Christ. The tree has often been used in Europe to form hedgerows and fences, since its thorns discouraged pigs and other animals from going through. The wood of the hawthorn is very hard, and the scientific name *Crataegus* comes from the Greek *kratos*, meaning strong. It is used in making walking sticks. In Missouri, the hawthorn is the state tree.

hazelnuts Three species of hazelnut grow wild in North America. These include the American hazelnut, *Corylus americana*, which grows in the eastern part of the United States, north into Canada, south to Tennessee, and west to the Dakotas. The beaked hazelnut covers much of the same range, and reaches over to the Pacific. Still another species grows wild in the mountains of California. The hazelnut is one of the better wild foods, free for the eating during late summer and autumn. The nuts can be shelled and eaten out of hand, used like any other nut in candies and sweets, and ground into a flour for making breadstuffs.

Several other species of *Corylus* grow wild in the north temperate regions of the Northern Hemisphere, ranging in size from shrubs to large trees 120 feet high. Several species and varieties are cultivated, and the nuts are usually marketed under the name filberts. Some species are also called cobnuts, Barcelona nuts, and Turkish nuts.
(See also NUTS.)

heading vegetables Admittedly, this is a catchall heading meant to include common-fare vegetables such as cabbage, Brussels sprouts, broccoli, and cauliflower, all of which are well known and widely available. Actually, all of these are varieties of the same plant, *Brassica oleracea*.

Lettuce (not all of which form as heads) has been cultivated for a long time, going back at least to 800 B.C., when it was raised in the gardens of Babylon. But it was the Romans who taught it to form heads. Another "head" vegetable, the artichoke, was probably raised first for its leaves. The globe artichoke is probably of fairly recent origin (in Renaissance Italy from native cardoons), but food historians dispute this point. A number of other vegetables such as mustard and capers also develop small headlike parts, or buds, that can be eaten.

hemlocks Not to be confused with water hemlock, which is highly poisonous, these trees of the pine family comprise four species that grow in the temperate regions of North America and six in Asia. The American Indians used the bark and the dried sap of the hemlock tree for making bread. For survival purposes, the bark can also be eaten raw or boiled. The Iroquois used the leaves of the Canada or eastern hemlock (*Tsuga canadensis*) for brewing tea, which is high in vitamin C.

henbits (*Lamium amplexicaule*) Native to Eurasia, these small annual herbs, members of the mint family that sport tiny purple flowers, now grow wild in most parts of North America. In spring, the young leaves and shoots can be gathered and eaten raw in salads or cooked as potherbs.
(See also HERBS AND SPICES.)

herbs and spices Generally, these plants and plant parts are used in small amounts to flavor larger quantities of other foods. The list of such herbs and spices is quite long, with many local favorites. Often, such an herb or spice may have been popular at one period of history, but may have been more or less ignored by earlier or later generations. Some plants may also be used, in some areas, to such an extent that they border on being classified as vegetables instead of herbs, whereas in other areas the same plants are used mostly for flavor. For example, cilantro is widely eaten—root, stem, and leaf—in Southeast Asia, whereas in the United States it is used sparingly (if at all), rather like parsley. Also, the seeds of cilantro have been used since ancient times as the spice coriander, which is marketed as a seed and as powder and is also used in curry powder and chili powder. Many herbs are used for color or garnish as well as flavor.

Spices and herbs have been used since ancient times, and have been important in the course of human history. Many trade routes were established because of the spice trade, and wars have been fought because of the profit to be made. For example, Columbus's voyage to America was inspired and financed by the interest in finding a shorter water route from Europe to the Far East in order to obtain spices.

The significance of particular spices tends to vary with the times and cultures, but two species of plants bearing the same common name have had long-standing influence on the various cuisines of the world. The most important is black pepper (*Piper nigrum*), native to Burma and Assam, which transcended its use in cooking and became a medium of exchange and even ransom. At one time, black pepper was more valuable than gold in international exchange. The spice is made from the ripe berries (peppercorns) of the plant, and white pepper is made from unripe berries of the same plant. Although there are a number of species and varieties, by far the most significant was, and is, *Piper nigrum*.

The other important pepper plant is the chili, including many species of genus *Capsicum*. There must be 1,000 species and varieties, ranging from the 1-inch-long heat capsules to the large, comparatively bland green peppers or bell peppers. The *Capsicum* species, used in dried, flaked, and powdered form, include hot red pepper or cayenne and the mild paprika. The *Capsicum* species also provide a hot juice that is used in pepper

sauces, such as tabasco. In any case, *Capsicum* peppers are native to the New World, and are now used in Africa, Asia, and Europe. Almost everywhere they have greatly influenced the native cuisine, especially in Africa and parts of Asia. The term pepper, as applied to the chili, apparently came from Christopher Columbus, who mistook powdered chili for a variety of *Piper nigrum*.

(See also ANGELICA; BREADROOT; BUGLEWEED; CILANTRO; FEIJOAS; GLASSWORTS; GOOD KING HENRIES; HENBITS; PEPPERGRASSES.)

hickories and pecans (genus *Carya*) There are about two dozen species of these nut-bearing trees. Three grow in Southeast Asia, and the rest are native to North America, mostly in the eastern states. Fossil remains of *Carya* have been found, however, in the Far North, including Alaska and Iceland. As a group, hickory nuts are quite tasty, but they have hard shells and small meats fitted into many crevices, making them difficult to shell. The wild pecan has a thinner shell and larger meats, and a "papershell" variety has been developed for commercial use, along with many other varieties.

In any case, most of these nuts found in the wild are edible, but a couple of species contain much tannin and have a bitter taste. In addition to bearing edible nuts, most of the hickory trees produce a sweet sap that can be used to make a syrup.

The wood of hickory is hard and tough, and is of value for handles for hoes, axes, and other tools. It is also a good fuel, and in the American South it is widely used for smoking meats.

(See also NUTS.)

highbush cranberries (*Viburnum trilobum*) These shrubs—which are not true cranberries—grow in moist places along the Canadian border, roughly 200 miles to the north and 200 miles to the south of the line, from the Atlantic to the Pacific. Then they turn north along the west coast and go on into Alaska. A related species, *Viburnum pauciflorum*, grows in the Pacific Northwest, and other similar species with edible berries are called hobblebush, squashberry, and nannyberry.

The highbush cranberry grows up to 16 feet high and produces small bunches of red berries. Because the berries tend to hang on the tree for a long time, they make an excellent survival food; but be warned that the berries are considered to be more or less poisonous if they are eaten raw. The fruit is quite similar to that of the cranberry, except that it has a large, flattened seed that must be discarded before the fruit's pulp is eaten. Highbush cranberries make an excellent pie as well as sauce and drinks. They are a "sweet and sour" fruit, which makes them ideal for making jelly. Bradford Angier, author of *Free for the Eating*, declares that there's no other jelly quite as good.

High in vitamin C, the plant is a good cure for scurvy. The American pioneers used a tea made from the bush's bark as a remedy for cramps. (In fact, the bush is sometimes called cramp bark.) The Indians of the Northeast used the bark tea as a treatment for mumps. The Indians also made a jelly with the berries and maple sap, and they smoked the bark as a substitute for tobacco.

(See also CRANBERRIES AND BLUE BERRIES.)

hops The common hop vine, *Humulus lupulus* and several related species, grow wild in North America, Europe, Africa, and Asia, and they are extensively cultivated for their small "cones" or "hops." Used in brewing for at least 1,200 years, these hops give beer its characteristic bitter taste and aroma. Hops also help preserve the beer. After being boiled in the beer wort, the hops are sometimes dried and sold as hops manure. The hops contain lupulin, a drug with sedative and hypnotic properties. American Indians as well as Europeans (including the physicians of King George III) used hops to promote sleep. But hops have also been considered to be unwholesome in England, and their use was prohibited by Henry VI and Henry VIII.

Although hops are not often eaten by today's foragers—possibly because of the bitter taste—the shoots, which can grow at a rate of 6 inches per day in spring, are nonetheless edible. According to a report in Michael A. Weiner's *Earth Medicine-Earth Foods*, they were at one time eaten as a substitute for asparagus. Weiner quotes a herbal of 1633:

> The buds or first sprouts which come forth in the spring used to be eaten in salads; yet they are, as Pliny saith, more toothsome than nourishing, for they yield but very small nourishment. . . . Only the young shoots are tasty, the older ones being so bitter and tough that some people bleached them with sulphuric oxide to soften them. King Henry VIII feared that he would be poisoned by this violent bleaching agent, an early food "softener," and he protected himself by passing an edict that forbade the addition of hops to ale brewed in his household.

The oil of hops is sometimes used in perfume, and the stems have been used for manufacturing fiber. The vine grows out anew each spring, and can reach 25 feet at maturity. It's an entwining vine, and always winds clockwise. In commercial production, the young stems are trained to climb strings hanging down from wire trellises.

I

Iceland mosses (*Cetraria islandica*)
These lichens grow to about 4 inches high in the arctic regions of the Northern Hemisphere and in some mountain ranges, including the Alps of Europe and the Appalachians in the United States, where they venture as far south as Tennessee. They are common on the lava slopes of Iceland.

Iceland moss contains large amounts of starch (lichenin) and, when boiled, it forms a gelatinlike substance or extract. This substance is sometimes served as food, especially in northern areas where other foods are scarce and where Iceland moss grows well. The extract is sometimes flavored with lemon or wine, or sweetened with sugar like jello.

Other lichens are also used for food. (See also REINDEER MOSSES.)

Indian cucumbers (*Medeola virginiana*)
These plants, native to North America, grow in wet woodlands from Nova Scotia to Florida, west to the Mississippi River. They are perennial herbs with a tuberlike rootstalk that tastes like a raw potato with a mild cucumber flavor. They can be peeled and eaten raw as a trail snack or in a salad, or they can be pickled. The scientific name *Medeola* is from the mythological Greek sorceress Medea. The plants are also known as cushatlilies.

(See also ROOT AND TUBER VEGETABLES.)

J

jackfruit (*Artocarpus integrifolia*) These tropical fruits are often confused with breadfruits. They can be eaten raw like fruits or cooked as vegetables, but they are not generally as well received culinarily as breadfruits and they have a disagreeable odor when first cut. Each fruit is hearty, weighing up to 70 pounds. The seeds are also edible.

(See also BREAD FRUITS; TROPICAL FRUITS.)

jack-in-the-pulpits (*Arisaema triphyllum*) Also called Indian turnips, bog onions, and starchworts, these unusual plants grow in North America from Nova Scotia south to Florida and west to Minnesota and Texas. The plant form a hooded spathe, which helps trap insects. The undergrown corm or bulblike root can be toxic. Nonetheless, some American Indian tribes routinely ate the corms after first drying them for a suitable length of time, which removes their toxicity. Modern foragers recommend that the plants be sliced thinly and dried. After proper drying, the slices can be eaten raw—perhaps with dip, much like potato chips—or the dried corm slices can be ground into flour and made into pancakes and breadstuffs.

Some sources say that jack-in-the-pulpit should be boiled in several changes of water but, apparently, it is the drying process that makes the plant edible. Although the corm can be dangerous, it is not likely that anyone would eat enough of the uncured corms to prove fatal. The taste sensation has been described as that of many needles pricking

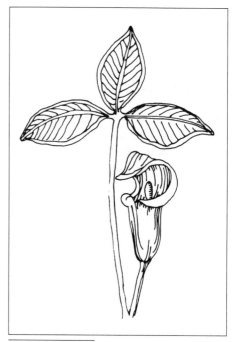

Jack-in-the-pulpit

away at the tongue. This sensation is somewhat relieved if the victim holds perfectly still. In fact, some Indian tribes require their young braves, as a test of manhood, to eat a fresh jack-in-the-pulpit corm without flinching.

Japanese knotweeds (*Polygonum cuspidafum*) One of some 200 species of knotweed that grow in the temperate regions of the world, Japanese knotweeds were introduced into the northeastern part of the United States from Asia. The young shoots are picked in spring and used as a substitute for asparagus. When older, the stalks are peeled and used like rhubarb for making pies and jams.

Jerusalem artichokes (*Helianthus tuberosus*) These plants have no relation to the artichoke and have nothing to do

Jerusalem artichoke

with Jerusalem, although they are now cultivated for food in the Holy Land. They are American sunflowers that have large edible tubers, which look like irregular and knotty potatoes or sweet potatoes and taste somewhat like an artichoke. In the wild, they grow along fencerows and other open places as well as in some dry woodlands. Their natural range is from Florida to Ontario and Manitoba. (In France, they are sometimes called *artichauts du Canada*.) Jerusalem artichokes have been widely transplanted and cultivated, and may be found in the wild far outside their original range.

When the Europeans first came to America, the Indians were cultivating the Jerusalem artichoke for food; the plant quickly became important to the settlers, and eventually gained popularity in Europe and around the Mediterranean. The Spanish called it *girasol*; the Italians, *girasole*. For a while it became more popular in London than the potato (another American import); somehow the English speaking farmers translated the Spanish or Italian name as "Jerusalem." The other part of the name probably came about because the tuber is similar to the artichoke in taste and texture, and can be used in the same recipes.

The tubers grow in the summer and can be dug in fall winter, and sometimes on into the spring. The Jerusalem artichoke's long availability, along with its good taste, make it a favorite among foragers. The American Indians ate the tuber raw, boiled, or baked. The tubers can also be sliced and fried, or peeled and pickled. They make a tasty variation of potato salad. The tubers are very low in starch, which makes them a valuable addition to some diets. The plants are sometimes found at markets, and home gardeners often find a spot for the plants along a fence or border. (Seed tubers are available at some retail garden supply stores and catalogs.) They make bright yellow flowers in addition to the tubers, which grow up to 5 inches long. The tubers can be harvested in the fall or early winter, and the surplus, if any, can be packaged and frozen. In some retail vegetable outlets, the tubers are available throughout the year, and are sometimes marketed under the commercial name sun root. It's not a bad name, especially since the tuber is the root of a sunflower. In fact, the American Indians called it sunroot.

(See also ROOT AND TUBER VEGETABLES; SUNFLOWERS.)

jewelweeds (*Impatiens biflora and I. pallida*) These two plants, both edible, are also called touch-me-nots. The term jewelweeds stems from the fact that in early morning the dew glistens on their drooping, waxy leaves like jewels. The term touch-me-nots comes about because the plants produce an elongated fruit, somewhat like a capsule, containing tiny seeds that are scattered when the fruit bursts open. This "explosion" often takes place when the fruits are touched by man.

In any case, the young shoots can be cooked like potherbs, but be warned that some authorities caution that the greens should not be eaten frequently. Also, the seeds can be harvested and eaten as nuts.

Both species of jewelweed are native to the eastern part of the United States and lower Canada, south almost to Florida. The plants have been introduced to Europe and other areas, where they have been naturalized. Darwin found the plants interesting because in America they depend on hummingbirds for pollination; in England, however, they reproduced without the help of the hummingbird. It was found that they also produce "blind" flowers that assist in self-fertilization of the seeds.

The Potawatomi Indians used juice of jewelweed to treat skin disorders, and to relieve the itch caused by poison ivy. Other tribes also used the plant for medicinal purposes.

Jew's ears (*Auricularia auricula*) These edible fungi grow on stumps and decaying wood throughout the world's temperate zones. Typically, they are tough and somewhat gelatinous or jellylike. They are often dried before eating, however, in which case they are hard and must be reconstituted in water before they are cooked. This species (or perhaps *Auricularia polyrichia*, which is cultivated in China) is marketed in dried form in Asian food markets.

Both species are called Jew's ear, cloudear, or woodear and, in fact, they

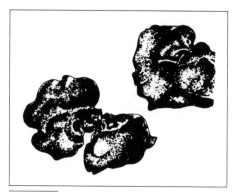

Jew's ears

resemble a human ear. According to Waverley Root, the popular term Jew's ear is a corruption of Judas's ear because, he says, the fungus grows on the elder, which was believed by some people to be the tree on which Judas hanged himself.

Several other species of fungi (or mushrooms) grow on dead trees, and are relatively easy for the forager to spot. These include the tree oyster, the beefsteak mushroom, and others. Anyone interested in eating wild mushrooms and fungi should consult one of the many field guides and manuals available on the subject.

(See also MUSHROOMS.)

jícamas (*Pachyrhizus erosus*) These vines, also called yam bean or Mexican yam bean vines, produce a rather large edible tuber as well as bean pods. The pods and beans are poisonous when mature, but the immature pods are sometimes cooked and eaten. The vines and roots are also poisonous. The tuber, when peeled, can be eaten raw or cooked. It has a pleasant taste that goes with other vegetables and fruits, and is thus a happy addition to salads. It is crunchy, like water chestnuts.

The jícama is a native of Mexico, where it is used in traditional cookery. A similar species (*Pachyrhizus tuberosus*) originated in South America. The Spanish transplanted the vine to the Philippines, and the Chinese took it to mainland China. These days jícama is used in Asian as well as Mexican and South American cookery, and it is available in many Asian and Latin American markets in the United States, as well as in some supermarkets. It takes several shapes, but usually looks rather like a mature turnip root. The tubers grow up to 6 pounds, although much smaller ones are often marketed. When fresh, the jícama keeps well and has cream-colored flesh that is juicy and crunchy.

(See also ROOT AND TUBER VEGETABLES.)

Judas trees These trees burst forth in early spring, around the time of Easter, with many red buds, long before the green leaves start to appear. In America they are often called redbuds. Several species grow around the world, including the Eurasian *Cercis siliquastrum* and the American *Cercis canadenis*, which grows from Ontario to Mexico. According to a popular belief, the redbud (also, some people believe it was the elder) is the tree from which Judas Iscariot hanged himself.

The buds and flowers are edible, either raw in a mixed salad or in such cooked dishes as fritters. The young pods (as well as the flowers and buds) can be stir-fried, like snow peas, along with other vegetables. The dried flowers also make a good tea. Both the pods and the buds can also be pickled. Within its range in America, the redbud grows wild and is often planted or transplanted as a decorative tree.

jujubes A member of the buckthorn family, the common jujube (*Zizyphus jujuba*) has been cultivated in China for over 4,000 years. Although it grows in other parts of Asia, it probably originated in China, where even today it is an important fruit. Outside China, it is often called the Chinese date; inside China, it is often called the red date. Although it's not a date, it is of the same size and shape, and has the same taste and sweetness.

Apart from its obvious advantages as a fruit per se, the jujube is a valued ingredient in Chinese cuisine. It is used to flavor soups as well as meat and fish dishes. During the time of Confucius, roast suckling pig stuffed with jujube was a culinary favorite. At one time (according to Marco Polo) it was used for making flour in China and, in fact, it is still used for that purpose in parts of Africa. The fruit has also been used to make a spirited drink.

In addition to the common jujube, several other species are grown in India and elsewhere. One species, *Zizyphus obtusitulia*, called the Texas buckthorn, grows in the American Southwest. The common jujube, however, yields the largest and best fruit. In addition to being used extensively in China, the small, thorny trees (which grow up to 30 feet high) are also cultivated in other parts of the world. It has been grown to a limited extent in Florida and California, but has not gained much of a foothold in the American market. In the Southwest, it is sometimes seen in home gardens and yards.

The jujube grows wild in Tunisia, and some scholars believe that the Tunisian isle of Djerba was the Land of the Lotus Eaters, where Ulysses and his men stayed, content to eat of the jujube and, no doubt, to drink of its spirits.

juneberries (genus *Amelanchier*) About 25 species of these shrubs and small trees grow in Eurasia, North Africa, and North America. They break out with white blooms in early spring and put forth many fruits, which are ready to eat by late summer. The fruits have 10 seeds, and these add to the flavor, especially when cooked. Juneberries can be eaten out of hand, or used in pies and jellies. In general, they resemble blueberries and can be used in the same way.

The American Indians, trappers, and early explorers made extensive use of juneberries, either fresh or dried. Often they were used in pemmican. The fruits can also be canned successfully. Some people believe that dried or canned juneberries have a better flavor than fresh ones. Of course, some species of juneberries taste better than others, and *Amelanchier canadenis* of the Northeast is a favorite.

The juneberries are often called serviceberries or shadberries. According the *Larousse Gastronomique*, they grow freely in the mountainous regions of Europe and along the banks of the Seine.

junipers About 40 species of juniper tree grow around the Northern Hemisphere, and the common juniper (*Juniperus communis*) grows widely in North America, Asia, and Europe, even at great elevations. In Lapland, the bark of a juniper is used in making rope. The oil of juniper has been used as a drug to promote urination, and the Paiute Indians drank a tonic made of juniper twigs to treat syphilis. The berries of some species are used, and marketed, as a spice, which is often used to flavor venison and other meats. The taste of juniper berries resembles that of sage.

The inner bark of the tree is edible and the berries have been used as a survival food. Some of the American Indians dried the berries for use in breadstuff throughout the year. The dried berries have also been used as a coffee substitute. As food, the berries are high in sugar but have a resinous flavor. The California juniper (*Juniperus californica*) has relatively mild berries and was widely eaten by the Indians of the region. In Asia Minor, *Juniperus drupacea* has rather large edible fruit known as *habhel*.

In any case, the juniper berry is of minor importance as food to modern man. Its main use these days is to flavor gin.

K

kelps The largest of the seaweeds, kelps grow around the world, and one species or another has been important in the past as a source of potassium and iodine, obtained by first drying the seaweed and then burning it to obtain the ash. In modern times, various products from kelp have been used in the food industry; a couple of these include gelatin and alginates, which have been used even in products such as ice cream.

Kelp, a brown algae, is eaten in Scotland and Ireland, usually after cooking it down to a gruel. Kelp has also been eaten along both the Atlantic and Pacific coasts of North America, and the Indians of the Pacific Northwest made extensive use of it, partly because it was readily available all year. In Alaska, a relish is made of kelp. Often the American natives ate the thick midrib of the larger kelps. The midrib has a sweet flavor, and is sometimes peeled, cut into slices, and pickled in vinegar, spices, and sugar. These pickles have been described as tasting like watermelon-rind preserves with a hint of the sea. Also, the midribs are hollow and were used as containers to store seal oil. (Remember that the Pacific kelp, *Macrocystis pyrifera*, grows up to 100 feet long.)

But the great kelp eaters of today are the Japanese. Although it is eaten fresh as a salad in coastal regions, kelp is more often dried, boiled, and compressed into a substance called *kombu*, which is available in powdered or sheet form (or shredded). The powder is widely used for seasoning, and the sheets are sometimes cut into strips for use in fish dishes and other culinary creations. *Kombu* has become quite impor-

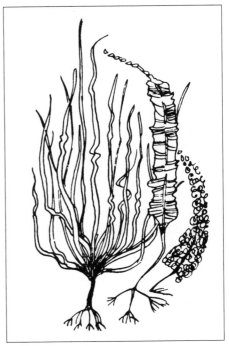

Kelp

tant in Japanese culture. In addition to being used in one form or another in almost every meal, *kombu* is eaten at wedding banquets, tea ceremonies, and New Year's celebrations.

In recent years, attempts have been made to make more extensive use of kelp as food for the world at large. But there are problems with it. Although kelp contains lots of iodine and potassium as well as vitamins A and D, it is low in protein.

Kerguelen's land cabbages (*Pringlea antiscorbutica*) This member of the cabbage family was discovered, probably in 1769, on the remote Island de Kerguelen (in the south Indian Ocean) by Captain James Cook, who made several exploratory voyages along the fringes of the antarctic and who was always very interested in feeding his ship's crew well. But the plant wasn't described in detail until after the Antarctic voyage made by Sir Joseph Dalton Hooker from 1839 until 1843. During this expedition, Hooker stayed for some time on Island de Kerguelen, and made daily use of the wild cabbagelike plant that grew there. It was eaten as a vegetable, and was also added to meat and pea soup.

The Island de Kerguelen is very far from any continent and is rather barren. At the time of Hooker's stay, it had only 18 flowering plants. Other members of the cabbage family (Cruciferae) are pollinated by insects; but there are no winged insects on the island and the plant had to adapt to wind pollination.

kudzus (*Pueraria lobata*) These rapidly climbing vines are used in Japan and China for making a useful fiber, and their starchy roots yield an edible starch. The name comes from a Japanese village, Kuzu, which grows a vine of exceptional quality. The plants grow extremely well in the southeastern United States, where they are sometimes planted for erosion control. Like many other introduced plants and animals (such as the carp), kudzu did more than it was intended to do. Here's the story, called "The Vine that Ate Dixie," reprinted from *A Cruise Guide to the Tennessee River, Tenn-Tom Waterway, and Lower Tombigbee River*:

> Even if you've never seen kudzu before, you'll have little trouble identifying it. This broad-leafed relative of the bean has taken over most of the available landscape throughout the South—enveloping trees, hills, power lines, sheds, junked cars, and the occasional slow moving politician [or farmer's slowly driven pickup trucks]. Kudzu is universally loathed and seldom mentioned without an accompanying string of epithets.
>
> It wasn't always so. Kudzu was grown successfully in the Orient for more than two thousand years for forage, as an ornamental plant, and for a pharmaceutical herb. Its tough fibers made durable cloth and paper, and it yielded a starchy flour substitute.
>
> When Kudzu was introduced at the New Orleans Exposition in 1884, agronomists thought it would be the saviour of the South's depleted farmlands, most of which had been "cottoned to death." Like other legumes, it can draw free nitrogen from the air, passing it to the soil through nitrogen- fixing bacteria on its roots and increasing the fertility of run-out soil without the use of expensive chemical fertilizers.
>
> The plant's impact on Southern culture was significant: Counties held annual kudzu festivals, and the Kudzu Club of America was formed in Atlanta in 1939 by Channing Cope. Dubbed the "Father of Kudzu," Cope urged farmers to plant the vine, proclaiming it the new king of Southern farming.
>
> Ironically, the vine's principal attraction proved to be its biggest drawback. Able to grow a foot or more *per day*, kudzu did indeed cover blighted landscapes in record time—and kept on going. . . .

L

lamb's quarters (*Chenopodium album*) These weeds, members of the goosefoot family, grow on agricultural lands the world over, except for very cold regions. The plants are often called pigweeds, apparently because they have been widely used for fodder as well as for human food. They were introduced into America from Europe, after which the plants were widely used by the American Indian.

By all accounts, lamb's quarters make an excellent green for eating. In the spring, the young leaves and shoot can be gathered and cooked like spinach or used in a green salad. Later, the young leaves in the top of the plant can be used in the same way, although the lower leaves will be quite bitter. The greens are high in vitamins A and C. As the plants mature, spikes develop and fill with tiny flowers and, finally, tiny seeds. As many as 75,000 seeds have been counted on a single plant. The seeds are eaten as a cereal, or they are dried and ground into a black flour, which is said to be highly nutritious. Reportedly, Napoleon lived on black bread made from seeds of lamb's quarters during a time of scarcity.

larch trees Members of the pine family, are unusual in that they shed their leaves in autumn, whereas most other pines are green all year. Several species of larch grow in the subarctic parts of America, Europe, and Asia, and the trees have been highly valued in the past for building ships and barges, and for use as poles and pilings. It has been said that the city of Venice stands on larch pilings. Also, a European larch yields a pitch that has been marketed as Venetian turpentine.

Although the larch tree is not noted for culinary uses, it has nevertheless been used as a source of food in North America and Siberia. The inner bark is edible either raw or pounded into a flour and cooked, and the Siberians use it to make a broth. The Ojibwa Indians drank a tea made from its roots, and the gum of the western larch (*Larix occidentalis*) has in the past been chewed for pleasure. In emergencies, the young shoots of the larch are eaten for survival.

latticeleaf plants (*Aponogeton fenestralis*) These aquatic plants of Madagascar have strange leaves from 6 to 18 inches long and 2 to 4 inches wide. The leaves spread out horizontally just beneath the surface of the water, and are so thin that they aren't much more than a lattice of veins, as the common name implies. The leaves are not important to the enthusiasts of wild food, but the tuberous roots are often eaten by the people of Madagascar.
(See also ROOT AND TUBER VEGETABLES.)

lavers These algae are usually red, but can sometimes be dark purple or purplish brown. A form of *Porphyra*, lavers are called sloke or sea spinach in Ireland, where they are often eaten. (The leaves resemble spinach.) They were also eaten by the American Indians on the Pacific Coast and Alaska, and are still consumed in large amounts by the Japanese and Chinese. Lavers are cultivated in Japan, and are usually marketed in dried form.

The plants can be eaten in a number of ways, including a pulverized form used in pancakes. The Irish simmer the plants for four or five hours and eat them with butter, cream, and lemon juice.

legumes Plants that produce seed pods are members of family Leguminosae, which comprises some 13,000 species of trees and smaller plants. The term legume is usually applied to edible species such as the various peas and beans, most of which have rather long pods containing a row of seeds. Legumes include peanuts (which are often called ground peas), soybeans, garden peas, field peas of various sorts, snap beans or green beans, snowpeas, and lentils. Most of these are eaten dried or fresh from the vine, and some are eaten in the pod. Black-eyed peas, for example, are usually considered to be dried fare, but many people in the southern part of the United States know that they are delicious when cooked fresh from the vine, shelled and boiled with a little pork seasoning. These shelled peas are often cooked with a few tender, immature pods, known as "snaps."

In any case, legumes of one sort or another are eaten in most parts of the world and they are highly important to man as food and sources of protein. Also, some of the widely cultivated forms, such as soybeans and peanuts, have nodules on their roots that help put nitrogen into the soil. These legumes are valuable in areas where crops are rotated so that the soil won't be depleted.

lichens See ICELAND MOSSES; REINDEER MOSSES.

licorices (*Glycyrrhiza lepidota*) The roots of this species of wild licorice were eaten by the American Indians, and are listed as survival fare in one manual. The Indians ate them raw, or roasted them in the embers of a fire. The word "licorice" is a corruption of *glycyrrhiza*, meaning sweet root. Indeed, the roots are used to manufacture a sweetening and flavoring agent (glucuronic acid, said to be many times sweeter than sugar). A similar species, called Spanish licorice, is cultivated for use in candies, tobacco, and other products.

Licorice has been used in various medicines for 3,000 years, and the American Indians used the roots to ease toothache. They also made earache drops by soaking the leaves in water.

litchis (*Litchi chinensis*) Also called lychees, lichees, leechees, lichis, laichees, not to mention litchi nuts, lychee nuts, and so on, these Chinese fruits have been a favorite in the Orient for centuries, especially esteemed in the Canton area of China. An old saying has it that once one tastes a litchi, he will never be satisfied until he gets another one. The fruit is cultivated for the market, and several varieties have been developed. The fruit does best in a warm climate that also has a cool or wet season, and it may not produce in years when the climate is less than ideal. The litchi is raised in Florida as well as in Hawaii, Mexico, South Africa, India, and, of course, in China, which leads the world in production.

About the size of a walnut, the litchi has a thin burrlike shell that slips right off the fruit, which is sweet but slightly tart. The grayish, translucent pulp is similar in texture to a peeled grape. Inside the pulp grows a seed, which is discarded. Fresh litchis are sometimes available in oriental markets (usually in summer). Canned and dried litchis are also marketed.

Lilies of the Nile

When the river rises and floods the plains, many lilies of a kind called by the Egyptians *lotus* grow in the water: these they gather and dry in the sun, and the centre of the lotus, like that of a poppy, they pound and bake into a bread. The root also of this lotus is eaten: it is round and as big as an apple, and it tastes somewhat sweet. There is another lily that grows in the river, like a rose, whose fruit is borne on a separate stalk coming from the root and is somewhat like a comb made by wasps. Inside it are seeds as big as olive-stones, and they can be eaten fresh or dried.

—Herodotus of Halicarnassus

The *rambutan*, a native of Malaysia, is a fruit that is very similar to the litchi, but is not as well known.

lotus lilies Sometimes called spatter-docks, yellow water lilies, pond lilies, and water chinquapins, these plants grow in freshwater ponds and tidewaters of America, China, India, Japan, and other countries, and were once an important part of the lower Nile culture. In ancient Egypt, the root of the white lotus was eaten, boiled or grilled, but the pink lotus was a sacred plant and forbidden as food. The lotus symbolized the sun and Horus, ruler of the sky. Hindus as well as Buddhists also give the lotus lily divine status, and in general the flower has been significant in both Eastern and Western symbolism and art.

In China, the seeds of the lotus lily as well as the large rootstocks are eaten and the plant is an integral part of some regional foods. The plant is also widely used in Japanese cookery. (In Japan, eating the lotus flower was believed to promote fecundity.) Rootstocks are sometimes available commercially in oriental markets, either fresh or canned.

The American Indians also made good use of the seeds of the American lotus (*Nelumbo lutea*), which they ate either fresh or dried, whole or ground into flour. They often parched the seeds on hot flat rocks. The Indians also ate the rootstocks, which they dislodged with their toes while wading. The rootstocks were also loved by muskrats, and the Indians sometimes stole the food from muskrat dens, usually when the animals were also hunted for food. (Some of the

American lotuses

Indians left other food for the muskrats, such as corn.) The water lily rootstocks were eaten raw, baked, roasted, or boiled. They were also dried and ground into flour for making breadstuffs and porridge. The Klamath Indians thought highly of the water lily, and even performed ceremonial dances before the harvest began.

Even the stem and the leaf (lily pad) of the lotus can be boiled and eaten for food if it is harvested before the leaf has finished unfolding. In China, the leaves are also used to wrap pork and other meat prior to steaming it. The seeds can be heated like popcorn; they don't pop but do swell up.

The Lotus-Eaters that were encountered by Odysseus on his journey were probably mythical, but the myth is supported by the fact that a bush (*Zizyphus lotus*), a member of the buckhorn family, grows a large edible fruit that was eaten by poor people. The fruit contains a mealy substance that was used to make bread. It was also used to brew a fermented beverage, or wine, which was believed to promote forgetfulness. The nettle tree of North Africa and Asia has also been suggested as the source of the fruit that detained Odysseus's men.

(See also JUJUBES; NETTLE TREES; ROOT AND TUBER VEGETABLES.)

M

mangosteens (*Garcinia mangostana*) A favorite fruit of Southeast Asia, the mangosteen grows to about the size of an orange and has tough, reddish purple rinds. Inside grow segments of snow-white flesh—juicy, delicate, and sweet-sour. The trees reach a height of 35 feet and have leaves about 10 inches long.

The mangosteen has long been cultivated in the Philippines and has been established somewhat in Hawaii, but it doesn't thrive in most areas. Also, the fruit doesn't ship well, which unfortunately makes it an unlikely fruit for the world's supermarkets.

(See also TROPICAL FRUITS.)

manzanitas (*Arctostaphylos*) Several species of these plants grow in California, and all are kin to the bearberries, the huckleberries, the cranberries, and others. In general, manzanitas are crooked-growing shrubs, which are sometimes used for ornamentals, that range in height from 3 to 10 feet. The plants produce berries which, when young, look like tiny apples. In fact, *manzanita* is the Spanish for "little apple."

The ripe berries of the manzanita are edible, and the California Indians sun-dried them for later use. Often they were mixed with pulverized dried venison and bear or goose fat. This nourishing mixture was often carried on long journeys. Also, crushed manzanita berries were used to make a cider drink. Some modern epicures and patio chefs use manzanita wood for smoking meat.

maples The American Indians depended on the maple tree not only for syrup and sugar but also for other foods as well as beverages. The inner bark was dried and pounded into a meal for baking bread. After being separated from the pod, the mature seeds were first boiled and then roasted, and the green seeds were boiled like peas. The young seedlings were eaten in spring as a green salad, or dried for later use. In addition to making syrup and sugar, the sap was used as a cold beverage, or fermented into a kind of beer. As a sweetener, maple syrup was also used in spruce beer and teas.

Several species of maple tree will produce a sweet sap that can be made into syrup but the best—in terms of quality and production—is the sugar maple, *Acer saccharum*. It grows in the northeastern United States, up into Quebec and Ontario, west to the Dakotas and south to Tennessee. Maple syrup is, of course, made by boiling down lots of sap. It takes at least 30 gallons of sap to make a gallon of syrup; the exact ratio depends on the tree, the season, and so forth. The Indians and early settlers obtained the sap by scoring the tree bark, or drilling into the tree, thereby causing it to bleed when the sap started to rise in the early spring. (The best season for getting the sap will have warm days and freezing nights.) The Indians turned the sap into syrup by putting hot stones into wooden troughs or earthen vats of sap. They also obtained some syrup by leaving the sap out to freeze at night; the next morning, they chipped off the ice, which contained mostly water, thereby concentrating the syrup in the bottom of the container. Modern practitioners have machinery and large distillery units to tap and contain the sap.

Although maple syrup is still important in parts of the Northeast and Midwest, it was the Indians who depended on it as a critical part of their food supply and culture. The Iroquois even had a "Maple Thanksgiving" day to celebrate the coming of spring and the rising of sap. Maple syrup time was truly a festive occasion, especially for the Indian children, who enjoyed candy made by pouring hot syrup into snow. Further, the syrup itself was a significant part of the diet. According to Michael A. Weiner's *Earth Medicine-Earth Foods*:

> Maple sugar, or syrup, was mixed with pulverized corn and taken as a highly nutritious food on long journeys. It was sometimes used as a sauce which was cooked into roasting meats. The Iroquois carried maple syrup on journeys in empty quail and duck eggs. These eggs were probably the first "no-deposit-no-return" containers, but they were biodegradable and not a source of permanent litter.

As implied earlier, other maple trees, such as the red maple and silver maple, will also yield excellent syrup and sugar. Also, a sweetener can be obtained from the sap of birch, sugarberry, sychamore, and walnut trees.

mare's-tails (*Hippuris vulgaris*) Erect plants that grow at the edge of lakes and ponds, mare's-tails were appreciated as food by the ancient Romans, and the plants are still eaten in a few places. Mare's-tails have a creeping rootstock that sends up new growths. The edible young shoots have been compared to asparagus, and are usually eaten after being boiled, steamed, or pickled. The plants grow in the arctic and temperate regions of the Northern Hemisphere, as well as in the southern part of South America.

marsh mallows (*Althaea officinalis*) Native to Eurasia and North Africa, these members of the mallow family have been naturalized in the northeastern part of the United States and perhaps elsewhere. The plants do best in salt marshes or at the edge of brackish water. In Europe, the mucilaginous roots have been used in sweetmeats and in the original "marshmallow" confection. Today the plants are used mostly by foragers, who eat the young leaves in summer, pickle the flower buds, and stir-fry the sliced roots. A candy is made by boiling the roots, then adding sugar to the water and simmering until thick. Also, the water in which roots have been boiled can be used as a substitute for egg white in meringue recipes or it can be used as cough syrup.

In the past, the marsh mallow had several medicinal and culinary uses, but today it is no longer widely used for anything. Related species of *Althaea*, including the hollyhock and the rose mallow, have edible leaves.

marsh marigolds (*Caltha palustris*) Highly poisonous when raw, the leaves of these plants are sometimes eaten after being cooked. Usually, they are boiled in two changes of water. The plants contain helleborin, which is bitter as well as toxic. The young flower buds are also eaten, like capers, after first heating them in very hot water (but not actually boiling them) and then pickling them in a solution of vinegar and other ingredients. Note that the water used to cook the leaves or the pickles should be discarded instead of being consumed as pot liquor. In folk medicine, a drop of juice squeezed from the leaves was used to remove warts.

Also called cowslip and May blob, the marsh marigold grows in swamps and marshlands across Canada from Newfoundland on to Alaska, dipping down into the the Dakotas and the Carolinas. It sometimes grows in great profusion, its yellow flowers blanketing wet meadows. The plant is of European origin and was introduced to America from Europe. In England, it is sometimes called kingcup.

Another species of marsh marigold, *Caltha leptosepala*, has white or bluish flowers and grows in the Rocky Mountains from Alaska to New Mexico. It also used as a potherb, but is not recommended for use in raw salads. This plant is sometimes called elkslip because elk like to feed on it.

The petals of some marigolds of the calendula species are also eaten in salads, omelets, stews, and other dishes. The Dutch, for example, are fond of adding marigold petals to soups. Also, the young buds are sometimes used like capers.

marsh woundworts These plants, which are sometimes known as nettles because of their hairy stems and leaves, aren't normally listed in handbooks or field guides to edible wild plants. But *Magic and Medicine of Plants*, published by *Reader's Digest*, says that the marsh woundwort (*Stachys palustris*) has abundant tubers that can be dug in the fall of the year and eaten rather like Jerusalem artichokes—raw, boiled, baked, or pickled.

Native to Europe, the marsh woundwort now can be found from Newfoundland to Alaska and throughout most parts of the United States. As the name im-

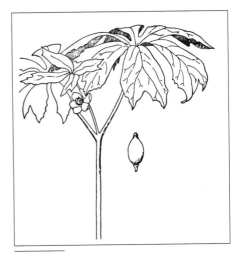

Mayapple

plies, it is found in marshes or wet places. The plants get the name woundwort because they have been used in folk medicine for treating sores and open wounds, and to stop bleeding.

(See also ROOT AND TUBER VEGETABLES.)

mayapples (*Podophyllum peltatum*) These poisonous North American plants of the barberry family have one or two large leaves atop a single stalk, making them look like umbrellas growing out of the ground. They range from Quebec to Florida, and west to Manitoba and Texas. About 20 inches tall, they sometimes cover large areas of ground. They bear edible fruits about the size of lemons. In fact, they are sometimes called wild lemons, as well as raccoon berries and mandrakes. The fruits ripen in the summer (usually July or August, depending on climate and geographical area), when they turn from green to a lemon-yellowish color.

When ripe, the fruits can be eaten out of hand, but they are usually used in jams and pies. The juice of the ripe fruits is also mixed with lemonade and other drinks.

Be warned that the leaves, stems, roots, and perhaps the unripen fruits of mayapples contain a powerful and posionous drug called podophyllin.

melons There has been some question as to whether melons are fruits or vegetables. In either case, they are members of the gourd family and there are two basic species. The mushmelon (*Cucumis melo*) originated in Iran or thereabouts, where it has long been cultivated. The larger watermelon (*Citrullus vulgaris*), which some authorities may not consider to be a true melon, originated in tropical Africa. Today there are a number of varieties of both types, including the honeydew melon, cantaloupe, and the ogen, a small melon named for a kibbutz in Israel.

Melons are usually eaten fresh, but they can also be dried. Marco Polo reported that the peoples of Shibarghan (in Afghanistan) sliced melons and dried them under the sun, whereupon they became "sweeter than honey." The seeds of some varieties are also eaten, usually dried and slightly parched. In addition, the rinds of some melons are used in preserves and pickles.

mesquites Several species of mesquite, both trees and shrubs, grow from South America to the southwestern part of the United States. The trees are seldom more than 50 feet tall, although their roots may grow as deep as 70 feet into the ground. Clearly, the mesquite is a plant designed to live in arid regions.

Some species, such as the glandular mesquite (*Prosopis glandulosa*) and the screwbean mesquite (*Prosopis pubescens*) have edible seedpods. When eaten green just before maturity, the pods have a pleasant lemon taste. The whole pod, including the pulp, is rich in sugar. When green, the pods can be cooked like string beans; when dry, they are ground into a powder and used in breads. Also, some Indians made both a sweet beverage and a beer from the pods. The dried seeds are sometimes ground and eaten in a mush. The tiny flowers can also be eaten, and are used for making a honeylike sweetening. Finally, the resinous gum that oozes out of the bark is also high in sugar and can be eaten raw or, better, boiled into a syrup.

Obviously, the mesquite was a valuable plant for the American Indians of the southwestern part of the United States and Mexico. It is still used as a source of food in some areas and, of course, it can be valuable as a survival food. But these days there seems to be more interest in mesquite wood chips for use as a smoking agent for backyard barbecues, and in charcoal fuel made from mesquite wood.

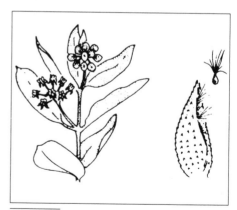

Milkweed

milkweeds (*Asclepias syriaca*) A native of North America, the common milkweed is also known as silkweed, cottonweed,

Virginia silk, and wild cotton. As these names suggest, the plant's large seed-pods produce a silky down that was used by the Indians and early settlers to stuff beds and pillows. The plant has even been cultivated like cotton and, during World War II, it was used in the manufacture of life jackets. Milkweed grows in the northeastern part of the United States and along the Canadian border into the Dakotas.

The name milkweed came from the fact that the plant has a milky white sap, which was used in folk medicine to cure everything from worms to warts. (Its scientific name, *Asclepias*, is derived from Asclepius, the name of the ancient Greek god of medicine.) Reportedly, the Indians made a chewing gum from this sap. The young shoots, buds, flowers, young pods, and new leaves can be cooked as vegetables, but they are usually boiled in two or more changes of water to remove the very bitter taste. Be warned that the roots of milkweed are more or less poisonous.

miner's lettuce (*Montia perfoliata*) Introduced from Europe, miner's lettuce grows on the west coast of America from Mexico to British Columbia, eastward to the Great Lakes. A cousin, Siberian miner's lettuce (*Montia sibirica*), grows in Alaska as well as in Siberia. The plants have an unusual shape in that the stems continue on through the center of the upper leaves.

The young stems and leaves of miner's lettuce, as well as the blossoms, are edible either raw as a salad or cooked as potherbs. The roots are also edible. Reportedly, the California Indians put bunches of miner's lettuce near the hole of a certain kind of ant. Then they cov-

Miner's lettuce

ered the hole and the plants, causing the ants to circulate through the plant, thereby imparting a sour vinegarlike flavor to the greens.

The early miners made extensive use of this plant (hence its name), which is high in vitamin C. It is sometimes called Indian lettuce or Spanish lettuce. In Europe it is referred to as winter purslane.

mormon tea (*Ephedra nevadensis*) These stunted shrubs of arid lands were used by the American and Mexican Indians to make a tea, and this use was picked up by the early settlers. The term "Mormon tea" probably became the popular name for the plant because the Mormons did not drink regular tea or coffee because of religious beliefs. Mormon tea is also called Brigham tea, desert tea, Mexican tea, squaw tea, joint fir, and joint pine. The tea was made by powdering the green twigs and stirring the powder into hot water.

The Indians believed that the tea had medicinal value for curing colds, fever, and headache. One account says that the Navajos burned the plant in a hole in the ground and squatted over it so that the

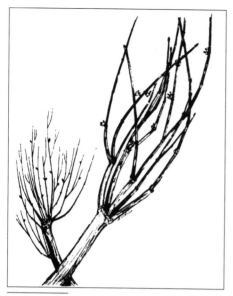

Mormon tea

smoke would touch their private parts, believing this to be a cure for venereal disease. Desert people also chewed the stems to sooth sunburned lips. In any case, the tea from the plant was consumed, and the Indians also ate the seeds of the plants either whole or ground into a flour.

Actually, several species very similar to the common *Ephedra nevadenis* were used in the American West. A Chinese species, *Ephedra sinica*, known as mahuang, has been used for thousands of years to treat asthma, improve blood circulation, relieve coughing, and reduce fevers. For a long time such claims were not taken seriously by Western scientists, but in recent years the plant was discovered to contain an alkaloid (named ephedrine) that is indeed useful in treating hay fever, asthma, low blood pressure, and other ailments. The common Mormon tea, however, is not known to contain ephedrine.

mountain ashes Members of the rose family, these trees and shrubs grow in the cooler areas of North America, Europe, and Asia. The American mountain ash (*Sorbus americana*) and the European mountain ash (*Sorbus aucuparia*) bear bunches of edible red berries. (A few other *Sorbus* species also have edible berries, but keep in mind that the so-called poison ash is really poison sumac, *Rhus vernix*.) The berries can be eaten out of hand in autumn and winter, when they tend to cling to the trees. They are, however, rather bitter until they are somewhat dried out. The berries can be dried and stored for future use or, as is often done, used in pies and jellies.

The European mountain ash was brought to America by the colonists, who used the berries like cranberries and for making wine. The berries have a high vitamin C content, and in folk medicine they have been used to treat hemorrhoids and diarrhea.

mountain sorrel (*Oxyrida digyna*) As its common name implies, mountain sorrel grows in high, rocky places, usually from 8,000 to 13,000 feet above sea level. In America, it grows in Greenland, across northern Canada, over much of Alaska, and south along the Rocky Mountains to New Mexico. It also grows in high places in Europe and Asia. Sometimes called alpine sorrel, the plant grows up to 2 feet in height and has a deep root. It has reddish flowers that grow on a stalk up above the green leaves.

The young leaves are edible and are at their best—tasty and juicy—before the flowers appear. However, the leaves have a slightly sour taste, and the plant is sometimes called sour grass. Consequently, the leaves are often mixed with other greens. The leaves are eaten raw in a salad or cooked. A tea is also made from the leaves, and the natives of the arctic regions of America and Siberia ferment the leaves like sauerkraut. The leaves can be put into chicken or beef broth to make a soup.

mulberries Twelve species of mulberries grow in the temperate regions of the Northern Hemisphere and in the mountains of some tropical regions. In addition to growing wild, some of the mulberries are cultivated for their edible fruit, for silkworm culture, and for other purposes. The largest is the North American red mulberry, *Morus rubra*, which grow up to 70 feet tall. It has red fruit that turns black when ripe, similar to a blackberrry in appearance. The fruit can be cooked in a variety of ways. It is sometimes eaten raw; but be warned that the unripe fruit and raw shoots contain hallucinogens. The white mulberry, a native of China, *Morus alba*, has white fruit and has been widely cultivated since the 12th century as food for silkworms in Asia and Europe. The fruits are eaten fresh or used in jams and preserves. The Afghans are said to mash the fruit together with almonds to make a paste. The white mulberry has also been cultivated in America, and it now grows wild from coast to coast. In addition to edible berries, which ripen in summer, both species have edible shoots in spring. These are picked just as the leaves are unfolding and are usually boiled and seasoned with butter.

The several other species include the black mulberry, *Morus nigra*, which is a native of western Asia. It was cultivated for its fruit by both the Greeks and the Romans, and it spread into northern Europe. It has an oblong purplish black fruit of superior quality. In general, the fruit of the wild mulberry is sweet but somewhat lacking in flavor. Special varieties have been developed specifically to produce better fruit, but these are usually grown for home consumption and not for market.

muscadines (*Vitis rotundifolia*) Often called bullaces or bullages, these large grapes grow wild in the southeastern United States, usually in thickly wooded creek bottoms and river valleys. The vines tend to reach far up into the trees, and the fruits are often high off the ground. As big around as a quarter and of a luscious purple color, muscadines are best when they ripen on the vine in August, September, or October, depending on the location and season. Muscadines can be eaten on the spot, or they can be gathered and used for making jelly, preserves, pies, juice, and even wine.

Muscadines and their cousins, the scuppernogs, can be grown at home or on the farm, usually on an arbor about 8 feet high and 12 feet square. Dozens of varieties are available for planting. Some are self-pollinating, while others require pollen from both male and female plants. Local gardening supply or seed stores, or mail-order houses, sell cuttings. Vines in the right spot will raise a bushel of fruit, or more. Some arbors have been known to produce for over 100 years at well-kempt homesteads. Scuppernogs, and even muscadines, can sometimes be purchased at supermarkets, roadside stands, and other markets. More and more, they are being grown commercially, especially in Mississippi.

The young leaves of wild muscadines are safe to eat as survival food. The long vines, when cut near the ground, will sometimes drip fresh drinking water that is quite pure.

(See also GRAPES.)

mushrooms Some 30,000 species of mushrooms grow in the world, often in dark places, and almost all of them are edible. That is to say, very few are poisonous. But a few species are deadly, and others are less poisonous so that eating a bite or two would not result in dangerous consequences whereas eating a bellyful might be quite serious. Of the many species that are not poisonous, a few are very good to eat, some are pretty good, and others are merely edible, being either bitter or tough as shoe leather, or both. At their best, mushrooms are a culinary delight, but the truth is that they are not very nutritious. They contain no sugar, few vitamins, not much carbohydrate, and less than 3 percent protein.

Mushrooms are cultivated by man—and by ants—in various parts of the world, and some interesting varieties are now available in specialized markets, either fresh or dried. There are some good cookbooks on modern mushroom cookery, and field guides to mushrooms are available for those who like to gather wild foods. Be warned that positive identification is recommended (in some cases a microscope is required) and be warned further that some folk tests are simply not reliable. Further, the "zones" listed in some books for the various mushrooms, poison or otherwise, may not always be entirely accurate or up-to-date. Deadly poisonous mushrooms have been found where they were not previously known to exist.

(See also JEW'S EARS; TRUFFLES.)

N

nasturtiums These common garden plants, comprising some 50 species, have a surprising culinary past. The Greek historian Xenophon said that the flowers were eaten by the Persians about 400 B.C., and that the plants spread to Turkey and other countries. In Paris, Alice B. Toklas, companion to Gertrude Stein, used nasturtiums in a salad that was, she said, based on a recipe from an old Turkish cookbook. In England, John Evelyn, a 17th-century man of letters, touted the nasturtium in his *Discourse on Sallets*. Dwight D. Eisenhower used nasturtiums in his recipe for vegetable soup.

The nasturtium was called Indian cress by the early American settlers, who found it growing wild. It has also been called Mexican cress, and even Jesuits' cress because it was apparently brought into Europe by the Jesuits. Like so many other edible plants, one species, *Tropaeolacae majus*, has been traced back to the mountains of Peru. Even today another species, called the anu, *Tropaeolacae tuberosum*, is raised in the Andes for its edible tubers. Usually, however, the tubers or roots are not eaten. (One source says that the whole plant can be eaten *except* the roots.) The leaves and blossoms are said to taste much like watercress, and they are usually eaten in salads or put into some other dish, such as omelets. The young green fruits are sometimes pickled in vinegar like capers. The plant is high in vitamin C.

For a dozen or so recipes for nasturtium, see *The Forgotten Art of Flower Cooking* by Leona Woodring Smith.

nettles (*Urtica dioica*) These edible plants are often called stinging nettles—and for good reason. They have stinging hairs, which contain small amounts of formic acid. But the sting is not too bad and doesn't last long for most people, although it can cause a rash. The plants are high in vitamins A and C, iron, and protein, and, fortunately, the hairs loose their sting when the plants are boiled for a few minutes. The leaves and shoots are best when the plant is young and less than 6 or 7 inches tall. Gloves and snippers are recommended for cutting nettle.

Several closely related species grow wild in North America and Eurasia. In the past, the fibers contained in the stalks of the stinging nettle were used to make a durable, linenlike cloth. In fact, this use of the plant is said to date back to the Bronze Age, and it has been suggested that the very name nettle means textile plant. Perhaps the word is kin to "net." In any case, the plant was at one time cultivated in Scotland for use in making cloth. The plant has also been used in folk medicine for the treatment of gout, baldness, and other ills.

nettle trees (*Celtis australis*) These trees of Eurasia and North Africa grow a small, sweet cherrylike fruit, which is first red and then black. The fruit is eaten in modern Africa. The tree is probably what the ancient Romans called the Libyan lotus and, as such, could be the plant that fed the Lotus-Eaters of Homer's *Odyssey*. The genus also comprises other species of trees, including the hackberry of North America.
(See also LOTUS LILIES.)

nuts Everybody knows about nuts, but there are a few surprises. A peanut, for example, is really a legume that puts on a pea under the ground instead of in the air. This explains the "pea" part of the name, and why they are sometimes called ground peas. Also, the litchi nut, sometimes called Chinese nut, is closer to a fruit than to a nut, at least to most people's way of thinking. In short, the term "nut" is not often precise, and all so-called nuts are not edible. Some are used for other purposes. For example, the tallow nut of China is not eaten but used to make wax for candles and soaps. The tung nut produces not food but an oil that is used to make paint. Many nuts have been used for making oil for use in cooking and foodstuffs, such as the coconut and the walnut.

As a rule, edible nuts contain a good deal of protein as well as fat. They have been quite important in times of famine to some people, but these days are not often used as a main food. Instead, they are used along with other foods (especially sweets, such as ice cream) or as snacks. In some cultures, various nuts are ground up and used in soups, stews, and other dishes. In America, peanut butter and jelly sandwiches have become something of a traditional food for hungry children to eat after school. In West Africa and Southeast Asia a peanut paste is frequently used as an essential ingredient in soups and stews and other main dishes. Nuts are also popular in some areas as stuffing for poultry.

(See also ACORNS; BEECH TREES; BRAZIL NUTS; BUTTERNUTS; CASHEWS; CHESTNUTS; GINKGO NUTS; HAZELNUTS; HICKORIES AND PECANS; LITCHIS; PINES; STONE FRUITS; WALNUTS.)

O

ocas (*Oxalis tuberosa*) Native to the Andes, these plants produce an edible tuber and have been cultivated at high altitudes (where potatoes won't grow) since ancient times. The tubers have an acid taste (*oxalis* is Greek for acid), which disappears after the tubers have been dried in the sun for 6 to 10 days.

According to *Larousse Gastronomique*, the oca was transplanted in 1829 from South America to England, where it grows nicely. The same source also says that the plant grows wild in the forests of France. It is also called occa, oka-plant, and oxalis. Another species, *Oxalis deppei*, also produces edible tubers in parts of Mexico.

(See also ROOT AND TUBER VEGETABLES.)

okra (*Hibiscus esculentus*) A native of Africa, this plant is now grown for food in the Middle East, India, South America, and the Caribbean. Brought to the southeastern part of the United States by the African slaves, okra is now a part of Cajun and southern cuisine. It is also an important ingredient in a popular Cajun dish called gumbo, to which it lends the distinctive mucilaginous quality. In fact, the word gumbo comes from the Angola *ngombo*, another name for okra. Any stew thickened with filé, or dried leaves of sassafrass, is also called gumbo, and these days the term is incorrectly applied—even in Cajun cookbooks—to just about any stew and especially to those containing seafood.

Only the immature pods (or fruits) of okra are eaten, and these are prepared by boiling, steaming, or frying. The pods are also pickled. In Africa and Madagascar, the pods are eaten fresh or dried, and the leaves of some varieties are also eaten—also fresh or dried—along with the young shoots. Today, okra is standard fare in parts of Africa and the Caribbean. It is known by several other names, including ladies' fingers, okro, quingombo, and bamia.

olives (*Olea europaea*) These important fruits might at first seem rather trivial to many of the world's people. Yet, they have been crucial to the commerce and culture of the peoples of the Mediterranean for centuries. The first olives used for food were probably picked wild somewhere in the Mediterranean or south-central Asia. They were cultivated on the island of Crete as early as 3500 B.C., and the importance of the olive goes back to Biblical history. According to the Old Testament, it was an olive leaf that the dove brought back to Noah's ark as a sign that the flood was over. The ancient Greeks prized the olive and its oil, and it was Plato's favorite food. The Romans cultivated it on a large scale, and they ate olives before a meal and after it; often they prepared dishes in which the olive was the main ingredient. The Romans also invented a screw press for extracting the oil from the fruit. Both the Greeks and the Romans used olive oil as a body lotion, as fuel for lamps, and to preserve food. (It is still used in canned tuna and other products.) The Egyptians also made use of olive oil, and may have used it as a lubricant to facilitate handling the large stones that were required to build the pyramids.

Somewhat surprisingly, the olive is not eaten in its natural state. Instead, it is pickled or canned before use. There

229

are several processes, but generally a green olive is one that is picked green and then pickled or cured; a black olive is one that is allowed to ripen on the tree before being picked and pickled or cured. Although most people in North America usually purchase olives in cans or bottles and consider them to be mere appetizers, they are sold by the pound or in other bulk measures in some parts of the world, and many varieties are seen in some Mediterranean markets. Olive oil is still an important commercial product, and over a million tons are marketed annually.

Spain is now a large commercial producer, and the plants are cultivated in all countries that border the Mediterranean. Olives are also raised in California, South America, and Australia. The trees live for a long time, and eventually grow into twisted, gnarled trunks. Some are at least 600 or 700 years old, and some are believed to be over 1,000 years old. In Jerusalem, some olive trees are said to have been living during the time of Christ and, according to Waverley Root, tourist guides in Athens point out the olive tree under which Plato sat while thinking great thoughts!

onions and garlic   (genus *Allium*) There are over 300 species of these plants, including leeks, shallots, and scallions. One kind or another has been cultivated for a long time, and the ancient Egyptians made extensive use of onions. Often, they were considered to be food for the common man, although rich folk also ate them either as a vegetable or as an agent to flavor other foods.

According to some experts, the Menominee Indians referred to a spot of land on the south shore of Lake Michigan as *shika'ko*, or Chicago, meaning "skunk place." The name did not come from the skunk as such but from the odor of wild leeks or onions that grew there in great numbers. In any case, about 70 species of onions are native to North America and they now grow wild just about everywhere. As a rule, the wild onions (including garlics and leeks) are a little strong to be consumed alone, although the entire plants—including the green stems—are nevertheless edible. They are good for flavoring meats and stews, for making onion soups, and for scrambling in eggs. The bulbs can also be pickled to advantage.

All species of *Allium* are probably safe to eat in moderation. These plants are easy to identify by their onionlike odor, but be warned that the death camass lily also looks like an onion.

(See also ROOT AND TUBER VEGETABLES.)

P

palms These trees grow mostly in the tropical regions, although some do thrive in temperate zones and one species survives 13,000 feet high in the Andes—2,000 feet above timberline. As a group, they have been important in man's history and development. All together, there are several thousand species of palm tree, some stately and some not much more than bushes. Most of the palms are concentrated in tropical Asia and tropical Central and South America. About 500 species grow in Brazil alone. Obviously, all of these trees can't be covered individually, but here are a few of special importance:

In the southern part of the United States, the sabal palm or palmetto (*Sabal palmetto*) was the source of the "swamp cabbage" that many early settlers enjoyed. During the Revolutionary War, a Polish commander in General Washington's army substituted swamp cabbage for real cabbage in *bigos*, a national dish of Poland. Swamp cabbage became a traditional dish for Florida hunting camps, and even today it is sometimes sold in swanky restaurants as heart of palm, where it brings such a high price that it is also called millionaire's salad. Sections of sabal palm were also sold at rural markets. Of course, only the inner core of the tree is used for this purpose. The trunk of this palm was at one time used as timber, especially for saltwater pilings, and the leaves are still used for thatching in some quaint structures. The palmetto is the state tree of South Carolina. The dwarf palmetto (*Sabel minor*) and several other species also live in the South, especially Florida, as well as in California and parts of the Southwest.

Sago is a yellowish flour or food starch that is produced from the trunk of several palm trees, and especially from the two sago palms (*Metroxylon sagu* and *Metroxylon rumphii*) that grow in the Indonesian archipelago. The economy of some of these islands depends on the sago exports, and sago is an important starch in the local diet. On the Moluccas Islands, sago, not rice, is the staple food. Sago is exported to France, Russia, and elsewhere.

The oil palm (*Elaeis guineensis*) grows wild in the tropical parts of western Africa and is now cultivated in Indonesia, the Malay archipelago, Central America, and South America. A similar species also grows in Africa. These trees bear a fruit that is up to 2 inches long and up to 1 inch in diameter. These fruits grow in large bunches, some of which weigh up to 35 pounds. When fully grown, a good tree can bear up to six bunches. Both the fibrous outer part of the fruit as well as the kernel bears oil. The kernel oil is used in margarine and other products, and the cake left from the kernels is used for cattle feed and fertilizer.

The betel or areca palm (*Areca catechu*) is native to Malaya, but is now cultivated in Ceylon, Thailand, and the Malay archipelago. The palms grow fruits, called betel nuts, which are widely used for chewing, just as chewing gum is used in the West. When not yet fully ripe, the fruits are picked, husked, boiled in water, cut into slices, and dried in the sun. Then they are wrapped, along with spices, in a small leaf from the betel pepper plant.

Many other palm trees around the world have edible cores or buds, and some bear edible fruit as well as nuts. The largest fruit or nut comes from the coco de mer (*Lodoicea sechellarum*) of the Seychelles Islands. The fruits of these palms weigh up to 40 pounds. Since they float in salt water, they were known in many parts of the world before their source was located.

The borassus or palmyra palm (*Borassus flabellifer*) is a favorite species in the south of India, Sri Lanka, and Burma. It is used for making a brown sugar, and the fermented sap makes toddy. The nuts are eaten, and are also used to make a drink.

In tropical America, the pejibaye palm (*Guilielma gasipaes*) produces an edible fruit.

Some palm leaves produce wax, and the collected sap of many trees is used for drinks of one sort or another, including wine, beer, and more ardent spirits. In India a palm beer is bottled commercially. Even sugar is produced and marketed from palms in India. Other commercial products from palms include timber, paper, starch, wax, tannin, and resin. The practical uses of palm trees by local peoples are too numerous to list, but they include fishnets made from palm leaves, basketry, fuel for fires, lookout towers for spotting fish in the lagoons of Pacific islands, thatching for roofs and so on.

Date palms and coconut palms (discussed in separate entries) have been, and still are, important sources of food for local consumption and for export. Life in the desert and on some remote islands would in the past have been almost impossible without these trees.

(See also COCONUT PALMS; DATE PALMS.)

papayas (*Carica papaya*) Popular as a breakfast fruit in the American tropics and the West Indies, the papaya is also used in salads, pies, and drinks. Within its natural range, it is said to be as popular as the apple is in North America. Apart from its considerable value as a food, it also contains papain, a substance used as a meat tenderizer and as an ingredient to aid digestion or in remedies for indigestion. In the tropics, native cooks wrap tough meat in papaya leaves for several hours to tenderize it. Meat is also tenderized by rubbing it with papaya juice. For commercial use, the papain is usually obtained from unripe fruit. Be warned, however, that the milky juice of unripe fruit can cause severe eye pain and even blindness.

The papaya fruits grow in clusters around the trunk of a palmlike tree, or shrub, that reaches a height of about 20 feet. The fruits grow just under the leaves, and the plant is quite profilic. In general, the fruit looks like a muskmelon, and has lots of edible seeds in the center. Depending on the variety, the shape of the fruit varies from round to bananalike, and some are pear shaped. The fruits vary in size as well as shape; some are rather small, others weigh up to 25 pounds.

The papaya is sometimes called the tree melon, and in parts of the West Indies it is called papaw or pawpaw. But it should not be confused with the wild pawpaw that grows in North America. In Cuba, the fruit is called *fruta bomba* and in the French islands it is referred to as *papaye*. Although the papaya is a fruit of the American tropics—it probably originated in Central America or perhaps in the West Indies—it is now grown commercially in suitable climates in other parts of the world, including southern Florida, southern California, and Hawaii. Unfortunately, the fruit has a thin skin and doesn't ship too well, unless it is picked quite green and hard. When ripe, the fruit softens to the touch and changes from a green color to a yellow or orange. Tree-ripened fruits have much more flavor.

In the tropics, the papaya is eaten cooked as well as raw. When small and immature, it is cooked like squash and, indeed, tastes rather like a summer squash. The green papaya is also used in relishes. The papaya is good source of vitamin C and potassium. For survival food, the young leaves, stems, and flowers are edible, but it's best to boil these in two changes of water.

(See also PAWPAWS; TROPICAL FRUITS.)

papyri (*Cyperus papyrus*) The ancient Egyptians once cultivated these shallow-water reeds, which were the basis for an important trade in an early form of writing paper. The town of Biblos—located on the coast of what is now Lebanon—was the center of international papyrus traffic, and from this town comes the word Bible. If papyrus had not been such a major product in ancient times, the Bible might well be called something else. During the heyday of papyrus, Biblos was an Egyptian dependency. Later it became an important city-state of Phoenicia. In addition to papyrus, it was the center of a trade in cedar—hence the Cedars of Lebanon.

In any case, the plant was vital commercially to early Egypt and, indeed, to civilization. Many important historical records were put down on papyrus. So many, in fact, that the science of papyrology was developed for the discovery, care, interpretation, and maintenance of papyri.

Papyrus was introduced to other parts of the world, and was actually cultivated in Sicily until more modern papers were developed. Although Assyrians called papyrus the reed of Egypt, it no longer exists in that country in the wild. It does grow wild in the upper Nile regions and in Ethiopia, and has been introduced in other places. Papyrus belongs to the sedge family, and is not a grass or a rush. (Some 7,000 species of sedge grow around the world, and some have edible roots or tubers, such as the chufa.) There is some confusion in ancient writings between the papyrus and the bulrush, and sometimes the words used in the Bible can be translated either way. The words "vessels of papyrus upon the water," for example, can and are also translated as "vessels of bulrush upon the water" (Isaiah 18:2). And it could well be that baby Moses was saved in a basket made of papyrus instead of in one made of bulrush. In any case, the papyrus reed was used in boats and sails as well as in mats and clothing. Its roots (which grew up to 10 cubits) were used as fuel and as wood for making various objects; it was, of course, also used in mummies. According to Herodotus, the pith of the reed was used as food, either cooked and raw. A favorite salad for the ancient Egyptians was made with lettuce and young papyrus shoots.

(See also BULRUSHES; CATTAILS; CHUFAS; REEDS.)

passionflowers (*Passiflora incarnata*) The name of these intricately beautiful American wildflowers is explained nicely in *Magic and Medicine of Plants*:

Exploring the New World in the 16th century, Spanish explorers were startled by the beauty of an exotic climbing vine whose white to pale lavender flowers seemed to symbolize the elements of the Crucifixion. The fringed corona, or crown, they felt, represented Christ's crown of thorns; the three stigmas, which receive the pollen, were the nails piercing the Savior's hands and feet; the five stamens were His wounds; the ten sepals and petals stood for the Apostles (leaving out Peter, who denied Christ, and Judas, who betrayed Him); and some said they saw the cross itself in the flower's center. The Spaniards accordingly named the plant passionflower. They also discovered that it grew throughout what is now the southeastern United States.

The passionflower is often called maypop. *Passiflora incarnata* bears a fruit that looks like a small melon, usually about the size of a large lemon. When ripe, these fruits are sometimes eaten out of hand in the southeastern United States, or they are used in jelly.

Actually, there are 350 species of *Passiflora*, most of which grow in South America. Some species are better than others for eating purposes, and *Passiflora edulis*, a native of Brazil, is grown commercially. It has, in fact, been cultivated in subtropical areas of India, Africa, Australia, and other places, including Florida and California. The fruits are sometimes available in American markets. The Houma Indians added the plant's pulverized roots to their drinking water as a tonic. Also, the dried fruits and flowers are used in herbal medicines.

(See also TROPICAL FRUITS.)

pawpaws (*Asimina triloba*) Native to North America, pawpaws are hunted by foragers in summer and autumn. Some-times called mayapples, pawpaws look like small apples, and the shrubs or small trees, growing from 10 to 50 feet high, are a member of the custard apple family (Annonaceae). The sweet yellowish pulp of the fruit, however, tastes more like a banana than like an apple, and, for unknown reasons, some trees bear better fruits than others. Pawpaws grow wild from New York to Florida, west to Nebraska and Texas, doing best in rich bottomlands.

Pawpaws can be picked while green and allowed to ripen in a dry place, but they are better when they ripen on the tree and fall to the ground. They turn from a yellowish green to a brown color. They can be peeled and eaten raw, but some people don't care for the raw fruit, and may even be allergic to it. Often the pawpaw is cooked in pies, ice cream, custards, and so on. It can be used as a substitute in recipes calling for banana, such as banana pudding. Some people like to bake the pawpaw in its skin, then peel and eat it.

Unfortunately, the tropical papaya is often called pawpaw, but the two are quite different fruits. Local names for the pawpaw also include Michigan banana and custard apple.

(See also PAPAYAS.)

pears See APPLES AND PEARS.

pecans See HICKORIES AND PECANS; NUTS.

peppergrasses (*Lepidium virginicum*) Although indigenous to Europe, these herbs now grow wild over most of the United States, the West Indies, and Mexico. They have also traveled to Australia. When young, the leaves, stems, and immature seedpods can be eaten raw in a salad or they can be cooked along with other greens. High in vitamins C and A, they impart a sharp, peppery flavor to salads, greens, and stews.

When the peppergrass plant is mature, the green seedpods can be stripped off and added to stews as herb seasonings. After the seeds dry in the fall, the pods can be stripped off the stems and winnowed, which will yield tiny, reddish seeds (two per pod). These seeds can be crushed and sprinkled on food like pepper, or they can be used whole in cooked dishes and in oil-and-vinegar salad dressings. A similar plant, called field cress or cow cress (*Lepidium campestre*), can be used in the same ways.

(See also HERBS AND SPICES.)

persimmons Known as kaki in China, Japan, and some other countries, the large oriental persimmon is now grown commercially and in home orchards around the Mediterranean and in the warmer regions of the United States and Chile. About 500 varieties have been developed, and they vary slightly in shape, color, and texture. In whatever form, the fruit is a delicious source of vitamin A, fiber, and potassium.

The name "persimmon," which came from the Algonquin language, was first applied to the much smaller American fruit, *Diospyros virginiana*, which grows wild and is a favorite of foragers within its range from Pennsylvania to the Gulf of Mexico and west to the plains. (Texas has its own species, *Diospyros texana*.) As compared to the large Chinese persimmon, the wild fruit is small (about 1½ inches in diameter) and has large seeds. The tree stands at a maximum of 40 feet.

The fruits are eaten raw, whether right off the tree or newly fallen on the ground, when fully ripe in the fall of the year. They are also used in pies, breads, and so forth, and can be dried for future use. Tea is sometimes made from persimmon leaves, and it is very high in vitamin C. Such tea has been used to prevent scurvy. Some Indians stripped the bark off the tree and boiled it in water, making a gargle solution that was used to ease mouth sores.

Also, the persimmon contains several large seeds, which are oval and flattened. The Indians ground these into meal for use in soups and breadstuffs. The American pioneers toasted the seeds and used them as a substitute for coffee. Survivalists might note that such seeds can often be found undigested in animal droppings near persimmon trees.

pickerelweeds (*Pontederia cordata*) These plants grow along the edges of ponds, lakes, and bays from the Atlantic Ocean to the Rocky Mountains. A sim-

Pickerelweed

ilar species, *Pontederia lanceolata*, has lance-shaped leaves and grows in the southeastern states. Both species have a beautiful flower stalk with many purple or blue flowers clustering on the end.

The young leafstalks, picked before the leaves unfurl, can be eaten raw in a salad or cooked as potherbs. The flower spikes develop fruits in the fall of the year, and each fruit contains a seed, which is edible as a snack. The seeds can also be dried and ground into a flour for breadstuffs.

pines The pine tree family comprises about 90 species that grow mostly in the north temperate zone; a few can be found in the tropics. Many of the pines have been, and still are, significant sources of timber and pulpwood. Others have been important as a source of food—pine seeds or nuts—and, indeed, such products are available commercially even today. The packaged nuts are sold in supermarkets in the United States, Europe, and the Middle East. Pine nuts (called *pignoli* in French cookery and confectionery) are often used in recipes.

The most important pine nut in America is the piñon, which is discussed separately. In the Mediterranean area, the seeds of the stone pine of Italy (*Pinus pinea*) are highly regarded and were enjoyed by the ancient Romans. *Pinus cembra*, a stone pine of Europe and Siberia, has oily seeds that are eaten as food, especially in the Alps and Siberia. This species is also used to make an oil, which is used for cooking and for lamplight. Most pine nuts are high in oil, and they are used in some diets as a substitute for meat.

The entry has concentrated so far on the pines that have rather large seeds that can be easily gathered. Most other pines have small or tiny seeds that can be eaten, especially in emergency situations. Squirrels often "cut" pinecones, and many birds feed on the seeds.

In addition to the seeds, the bark and needles of pines have been used as food, especially in times of need. The American Indians sometimes ate the inner bark or pounded it into a flour for making breadstuffs. The inner bark is full of sweet sap, especially in spring, and can be eaten fresh or dried for future use. The bark was an important source of food for some tribes, and the Mohawk name *Adirhon'dak*, meant "tree eaters." The pine needles, a source of vitamin C, were also used to make tea. Out west, the Indians made good use of the seeds of the piñon, the ponderosa, and the digger pine. A gum from the digger pine was used for chewing.

On the unusual side, the monkey puzzle (*Araucaria araucana*), a Chilean pine, grows to 150 feet tall and has stiff branches with dense sharp needles, making a tangled network that is difficult for monkeys and other animals to climb, hence, its name. The oblong cones reach a length of 5 inches or more and bear edible seeds.

In some cases, gathering pine nuts can be a community affair. According to Carleton S. Coon's *The Hunting Peoples*:

> In a certain section of southern Queensland in Australia there grows a pine, *Araucaria bidwillii*, that bears the so-called bunya-bunya fruit, only once in three years. It is a seed about two inches long that tastes like a roasted chestnut. These pine trees are privately owned, but they grow close enough together so that when the fruit is ripe their owners and their families used to assemble there. Some five to six hundred persons would come there to eat

the fruit as long as it lasted, which might be for several months.

(See also NUTS; PIÑONS.)

piñons (*Pinus monophylla.*) In some areas, these pine trees were once important in the diet and culture of the native peoples. Nowhere is this more evident than in the Zuñi culture of the southwestern United States and Mexico. (Note that other species of similar "piñon" pines have nuts in the American West, and they are marketed under the name "pine nut.") The Zuñi ate the nuts raw or parched. When dried, the nuts were sometimes ground into a meal, which was used for bread, sometimes mixed with cornmeal. The piñon was also mixed with seeds from yuccas and sunflowers.

In addition to the nuts, the pines provided green needles, which were high in vitamin C and used to make a tea. In hard times, the Zuñi also made a sort of sustenance bread from the bark of the piñon. The inner bark was boiled and pounded into a mash. Then it was shaped into pones or cakes and baked in an oven or pit. This bread was hard and tough, but it could be stored for months. Before being eaten, the bread was boiled in water to soften it.

Similar breads (and teas) were made by other American Indians from local pine trees.

plantains See BANANAS AND PLANTAINS.

plums (genus *Prunus*) These luscious fruits have been cultivated since ancient times, and many species and countless varieties have been developed. One of the more popular is the common European plum, *Prunus domestica*, which probably originated in the Caucasus and which is now cultivated in America and elsewhere. The damson plum may be even older, and it seems to have been cultivated around Damascus, Syria. These days, the damson is cultivated in California for use in making prunes. Other popular modern species include the Japanese plum, which is probably native to China but which was domesticated in Japan. Also, plums were domesticated by the ancient Incas.

Many other species are cultivated or grow wild, or both, on all of the continents except Antarctica. Several species grow wild in America, with the most popular species being *Prunus americana*. These are gathered in large numbers from small wild trees and eaten out of hand or used in pies, jellies, and sauces. Be warned that the large pits and seeds contain hydrocyanic acid, which, together with stomach acid, can lead to the internal production of deadly cyanin.

Other wild plums include the beach plum that grows along the coast of the northeastern United States, which is very popular for making jelly. The genus also includes the sloe (*Prunus spinosa*), which is not highly regarded as food but which is used in flavoring wines and liqueurs—especially sloe gin. Reportedly, unripe sloes have been sold in France as olives.

(See also STONE FRUITS.)

pokeweeds (*Phytolacca americana*) Sometimes called pokeberries or just plain pokes, these plants grow in the eastern part of the United States. Each spring wild food enthusiasts and foragers gather their young shoots and leaves just as they emerge from the ground. Pokeweeds are eaten like asparagus. The small leaves are also cooked like spinach or other greens. But the older leaves have a strong flavor and may not be safe to eat. The roots are definitely poisonous, and some people believe that the stalk and stems are too, after they turn purplish.

The berries of poke are also believed to be poisonous. These grow in bunches like fox grapes, and turn from green to purple. The berries have a red juice, and indeed the plant's name comes from the Indian word *pocan*, meaning a plant that yields a red dye. Poison or not, the Pamunkey Indians of Virginia drank a pokeberry tea as a cure for rheumatism. And, according to *The Foxfire Book*, poke root was roasted in ashes and then, while still hot, applied to a joint inflamed by rheumatism. The juice of the berry has also been used in folk medicine to treat hemorrhoids and cancer.

Pokeweed is sometimes cultivated for food, and has been transplanted to Europe. It can be raised for winter use by planting roots in a suitable container, exposing them to a hard freeze, and then moving them to a suitable place, such as a basement, with a sustained warm temperature. The shoots and young leaves are snipped off as soon as they are 3 or 4 inches high. Others follow.

pomegranates (*Punica granatum*) These fruits, the only members of family Punicaceae, have been important since biblical times, when King Solomon raised them in his orchard. The fruits probably originated in or around what is now Iran or Iraq, but spread to North Africa and eastward to India. They are now raised in many mild temperate regions, including the warmer parts of United States south to Peru. Pomegranates are an important part of the cuisine of some countries of the Middle East and the Caucasus, where they are used not only as fruit but also as flavoring for juices, marinades, sauces, and syrups, including the original grenadine that bartenders use to mix cocktails.

Pomegranates are popular with painters and poets as a symbol of fertility, owing no doubt to their many seeds.

poppies About 90 species of the poppy (family Papaveraceae, genus *Papaver*) grow in Europe, Asia, and North America. Opium is made from one species— *Papaver somniferum*—native to the Orient and Greece. It is cultivated for medical uses, and is grown or harvested illegally for use in the dope traffic. This same species is also used in the manufacture of poppy oil, which is used as salad oil, cooking oil, and so on. In the Ganges valley, poppy oil is burned in lamps. The seeds of poppy have been used in cooking for a long time, at least since 1500 B.C. in Egypt. They are still used for sprinkling on breads and cakes. Although slightly narcotic, the leaves are sometimes cooked and eaten like spinach, according to *Larousse Gastronomique*.

poverty weeds (*Monolepis nuttalliana*) Kin to lamb's quarters, these plants grow in the dry, alkaline areas between the Mississippi River and the Pacific, dipping up into Canada here and there. They also grow in central Alaska. Poverty weeds are annual herbs about 15 inches in height. The young leafy stems are edible and are usually boiled for a few minutes before being eaten with butter. The roots are also edible, but they do not receive high culinary marks. In the American Southwest, the Indians ground the seeds for use in breadstuffs.

prickly lettuces (*Lactuca scariola*) These common wild lettuces, probably fore-runners of the cultivated garden varieties, are native to Europe but now grow in most of the world's temperate regions. They shoot a stalk up to 7 feet high, then put out a number of stems bearing yellow flowers. The leaves have spines along the margins. When very young, the leaves can be eaten in a green salad or cooked as a potherb. The prickly lettuce has had a long history in herbal medicine, and it was believed to relieve pain. The plant is filled with a latex that hardens and turns brown when exposed to the air, looking and smelling like opium.

The prickly lettuce is also called horse thistle (because horses like to eat it), wild opium, and the compass plant. The latter name comes from the fact that the leaves of the prickly lettuce tend to turn to follow the sun during the day.

Several other species of *Lactuca* have edible leaves, including the wild lettuce, *Lactuca canadenis*, which has smooth leaves.

prickly pears Called Indian figs, several species of these fruit-bearing cacti, such as the common *Opuntia humifusa*, can be found in America and around the world. (During Operation Desert Storm, the prickly pear was listed as a survival food on a map issued to troops in the Persian Gulf area.) Most of the plants have padlike leaves, joined end to end, with terminal flowers and fruits. The flowers are usually yellow, but some are red and others white. The fruits can be black, red, green, or purple. The largest fruit comes from the *Opuntia megacantha* plant of Mexico and the southwestern United States. The various plants grow not only in the American southwest, but also in other areas, espe-cially sandy coastal regions. In all, there are 250 species of prickly pear, all of American origin.

Possibly because of the edible fig, the plants have been introduced to India, Africa, Hawaii, Australia, the Middle East, and other places. The plant took over parts of Australia to such an extent that insects and cactus diseases were in-troduced to control the plants. Prickly pears are raised for the fruit in Mexico, other parts of America, Africa, and southern Europe. They are also raised as food for the cochineal insect, which is used in the manufacture of dye. As many as 50,000 plants have been used on one cochineal farm.

The American Indians, of course, made wide use of the prickly pears. The Navajo Indians used forked sticks to pick the fruits off the plants, and the Apaches used wooden tongs. Usually, the fig is first peeled and then eaten raw. It can be dried for later use, and the Indians dried it and ground it with dried venison and fat, making a food of high nutritional value. The seeds of the fruits were also eaten, either roasted and ground into meal or dried for later use.

The green pads are also eaten. Usu-ally the Indians roasted these in the fire to remove the spines. They can also be peeled, sliced, boiled like string beans, or fried. The pads are tasty but somewhat slimy to the touch.

In addition, the prickly pear can be used as an emergency source of water, in which case the green pads are first peeled and then the pulp is chewed. The Black-foot Indians used the prickly pear to re-move warts. The pads, after being baked, were applied to wounds.

"Indian figs" are sometimes available in markets, and the pads of prickly pear or other form of cactus are also marketed, sometimes under the name *nopales*. These are important in modern Mexican cuisine.

(See also CACTI.)

purslanes (*Portulaca oleracea*) A native of India, purslane has been used as food for a very long time, especially in India and ancient Persia. The plant is still cultivated in home gardens in some parts of the world, and it grows wild over most of the tropic and temperate regions. Purslane prefers fields, fencerows, roadsides, pastures, and other rather open spots, and it can be found over most sections of both North and South America. In Mexico, it is a recognized part of everyday folk cooking, with recipes such as *Verdolagas con Chile Chipotle* (purslane with chipotle chile pepper). Unfortunately, it is not often used in the United States, except occasionally as feed for livestock and as human food by those people who enjoy wild foods. This is a great waste, since in summer and early fall purslane grows in great profusion, in bunches or mats, over many of our fields after corn and other crops have been harvested. It is one of the more widespread and one of the best wild greens, and is available after spring greens have grown too old to be palatable. It's easy to gather and easy to cook.

The bud, including the flowers and smaller leaves, is usually pinched off the plant. (Each cluster will have a number of buds, as limbs branch out from a central root.) These can be eaten raw in a salad, pickled, or cooked like other greens. The purslane is slightly mucilaginous, like okra, and is therefore a good ingredient for stews and especially for gumbo. The seeds of purslane can also be gathered, dried, and ground into a flour for use in pancakes and bread.

Purslane is also called pusley. Two other very similar species are edible: *Portulaca neglecta* and *Portulaca retusa*.

(See also GREENS.)

Q

quack grasses (*Agropyron repens*) Also called couch grass, dog grass, and witch-grass, quack grass, which shoots stems up to 4 feet high, grows wild over much of North America and Europe, to which it is indigenous. The plant has sometimes been planted on a limited basis as fodder for livestock, but this use is no longer common. These days it is usually considered to be a weed—and one that is hard to get rid of because it grows from adventurous rhizomes, which have many small roots at the nodes. Because of the network of roots, quack grass is sometimes used to control soil erosion.

Although the plant is kin to barley and has a scientific name that means "creeping field wheat," the part of the plant most often used for human food is the rhizome. In times of need, these have been dried and ground for use as flour to make bread. The rhizomes are also roasted and used as a substitute for coffee.

The plant is often called dog grass because dogs eat the flowering spikes and leaves when they have digestive problems. Also, herbalists sometimes use a tea made from the rhizomes because it is believed to solve urinary problems.

Queen Anne's lace (*Daucus carota*) Also called bee's nest, bird's nest, and devil's plague, the plant is more properly called wild carrot because it no doubt came from cultivated varieties of wild carrot. The common name, Queen Anne's lace, came about because the plant's stalk puts out large umbels of tiny white or pink flowers which, being intricate, suggest lace. These umbels also have a red or purple spot in the center, which corresponds to the fact that Queen Anne is said to have pricked her finger (drawing a drop of blood) while sewing lace.

The carrot is native to the area in and around Afghanistan, and it was brought to Europe as a vegetable to be cultivated; before long, it grew in the wild. The same thing happened in North America, where the wild plant is now quite common. Because the plant has been cultivated since before the Christian era, many forms and varieties exist. Most Americans think of carrots as being orange, but carrots of other colors are also raised, including white, yellow, and purple ones. The carrot is high in sugar (second only to the beet among root vegetables) and is sometimes used in puddings or other sweetstuffs by the Irish, the Hindus, and the Jews; a syruplike sweetening agent has also been made with carrot juice. Also, roasted carrot root has been used as a substitute for coffee.

In the wild, the carrot root is usually not as large or as good as the cultivated varieties, but it is edible if gathered while young. The young leaves can also be eaten in a green salad, or cooked in soups, and the flowerheads are sometimes dipped in batter and fried as fritters. The seeds can be dried and used as a seasoning.

At one time, carrots were used for medicinal purposes. The seeds were believed to eliminate intestinal worms and prevent flatulence. Also, the carrot, because of its high content of vitamin A, was believed to improve eyesight. Some books point out that vitamin A is harmful to humans if it is eaten in excessive amounts.

Be warned that wild carrots look quite similar to poison hemlock and water hemlock.

quickweeds (*Galinsoga parviflora*) Native to tropical America, these plants grow rapidly (thus their common name) and can now be found over most areas of North America. During World War II, a species of quickweed (*Galinsoga ciliata*) sprang up and covered London and parts of England almost overnight, apparently in areas that had been burned. Reportedly, the British, corrupting the scientific name *Galinsoga*, called it gallant soldier. In any case, quickweed is edible, and the tops are boiled as potherbs. Several closely related species are also edible.

quinces (*Cydonia oblonga*) A native or Persia or Iran—where it still grows wild as well as under cultivation—the quince has had a remarkable history. Some people even believe that the beautiful and fragrant fruit was the "apple" that was eaten by Eve, and it has a long reputation as being the fruit of love.

The quince was once a very popular fruit for making jellies because it contains lots of pectin, the stuff that causes "jelling." The Portuguese word for quince is *marmelo*—from which the English word marmalade comes. Perhaps the decline of the quince in America is due to the fact that home canning has declined in recent years.

In spite of it's decline in American cuisine, however, the quince is still going strong in the Middle East and other parts of the world, where it is often cooked with meats (especially lamb or mutton) and casseroles. In the Caucasus, for example, the quince is used in soups, meat dishes, compotes, desserts, and even confections, in addition to quince marmalades and preserves.

quinoa (*Chenopodium quinoa*) A member of the goosefoot family, the quinoa grows in South America on the Andean plateau of Chile, Peru, and Bolivia. It has been cultivated since ancient times. The green leaves are eaten as a potherb. More importantly, the quinoa produces a great many tiny seeds, and these are ground into flour and used in breadstuffs or gruel. The seeds are also cooked whole and eaten like rice, and are rich in calcium and vitamins.

Recently, the quinoa has been raised commercially in the Rocky Mountains, from New Mexico to parts of Canada. The grain is usually found in natural food stores, or gourmet shops, but is also showing up on the shelves of supermarkets. It's a healthy grain and may well find a place in the staples of the modern health-conscious world.

R

ramps (*Allium tricoccum*) These wild leeks are popular among foragers in the northeastern part of the United States, down the Appalachians to northern Georgia, west to Minnesota, and North into Canada. Ramps usually grow in shady places in rich woods, often under maple trees. The plant's young leaves (wide, like those of lilies) as well as the bulbs are edible. Usually, the leaves are added to salads, or used in soups, stews, scrambled eggs, and so on. The bulbs are eaten like other leeks or onions. The bulbs can also be pickled.

Gathering these wild leeks is a happy springtime event in some rural areas, and ramp festivals are held in Tennessee and in West Virginia. Some people love ramps, but others think they are too strong of flavor and aroma. As a character in the *Foxfire* books said, "They's not for ladies or those who court them."

(See also ROOT AND TUBER VEGETABLES.)

raspberries See BLACKBERRIES AND RASP-BERRIES.

reeds The common water reed, *Phragmites communis*, or reedgrass, grows over most of the United States and in many other parts of the world, from the Arctic to the tropics. It thrives in shallow water along the edges of lakes, sluggish streams, and marshes. Other species are also widespread. Since ancient times, reeds have been used for thatching, arrows, musical instruments, and writing pens. The American Indians made pipes from them, and used the flower plumes as diapers for newborn babies.

The reeds have also been an important source of food.

Reedgrass has stems that grow up from 5 to 15 feet above the waterline, and these are topped with beautiful plumelike flowers. The plumes themselves, which grow up to 16 inches long, are at first purplish, then tan and, finally, white. These bear reddish seeds that can be ground and used as flour. (Sometimes, however, the flowerhead fails to produce seed.) The young shoots and small leaves that stem from the long horizontal rootstock can be eaten raw as a salad or cooked like asparagus. The tender underground buds from the branching rootstock can also be eaten raw, but the roots themselves should be boiled to tenderize them. The roots can be dug all the year, making the reedgrass an excellent survival food.

(See also BULRUSHES; CATTAILS; PAPYRI.)

reindeer mosses (*Cladonia rangiferina*) These lichens are especially abundant in Canada and extremely common in most arctic regions. They can also be found in temperate zones, such as North Carolina. The lichens grow a few inches off the ground, and are sometimes compared to pasture grass. They are very important to reindeer and musk oxen.

Reindeer moss is eaten by man mostly as a survival food and, as such, it can be quite valuable for those in need. In Scandinavia, it has been used in the manufacture of alcohol, but the moss grows very slowly. When the land is stripped of its cover, 35 years are required to restore it to its former abundance. This is why the Lapps—who depend on the reindeer, which in turn depend on the moss—are nomadic.

The reindeer moss lifts right off the ground, much like clumps of purslane, and is thus easy to gather. It is available all year-round. The moss can be cooked and eaten in soups, or it can be dried and made into bread flour or mixed with potatoes. Sometimes it is soaked in a soda solution (or water containing campfire ashes) to remove the lichen acids—some of which can eat into rocks. The North American Indians boiled reindeer moss and ate it as a soup, or reduced it to a jelly.

A number of other lichens, including Iceland moss, are also eaten for survival. In North Africa and Asia Minor, the lichen *Lecanora esculenta* sometimes rolls across the desert, driven by the winds, and is said to be the manna of the Israelites. Various kinds of lichen have also been used in medicine and in dyes. The chemists' litmus paper, for example, depends on substances derived from lichens.

(See also ICELAND MOSS.)

root and tuber vegetables Everybody knows about such edible roots as carrots and onions, which are enlarged taproots. Other common types grow around the taproots, and are often enlarged portions of feeder roots. The latter group includes potatoes and Jerusalem artichokes. Many other kinds grow around the world, some of which—such as the taro, cassava, and yam—are extremely important food crops in some areas. In addition to cultivated plants, many wild plants also have edible tubers. These include the wild potato vine and the chufa.

There are a few surprises, however. A variety of parsley is cultivated for its roots as well as for its leaves. Also, celeriac—a variety of celery—is cultivated for its large, ugly root instead of its stalk. A good many vegetables have both edible roots or tubers and edible leaves. These include kohlrabi, turnips, radishes, and beets.

(See also AGAVES; AIR POTATOES; ARROW-ROOT; BITTERROOTS; BREADROOTS; BUNG-LEWEEDS; CAMASS LILIES; CASSAVAS; CHUFAS; GREENBRIERS; GROUNDNUTS; IN-DIAN CUCUMBERS; JERUSALEM ARTICHOKES; JÍCAMAS; LATTICELEAF PLANTS; LOTUS LIL-IES; MARSH WOUNDWORTS; ONIONS AND GARLIC; RAMPS; SALSIFIES; SEGO LILIES; SOW THISTLES; SPRING BEAUTIES; SWEET POTA-TOES; TAROS; TOOTHWORTS; WILD GINGER; WILD POTATO VINES; YAMS.)

roseroots (*Sedum rosea*) Growing on rocky ground across North America from Alaska to Newfoundland, these plants are easy to identify because of their edible root which, when scratched or crushed, smells like roses. Edible in spring and summer, the leaves and young stems are eaten raw in green salads or boiled as potherbs. The roots can be boiled and eaten, and are sometimes pickled. Roseroot is sometimes called "stonecrop" because it grows well in rocky places, and has also been called "scurvy grass" because it supplied lots of vitamin C to early explorers.

Roseroot is only one of 300 similar species of the *Sedum* genus that grow in the temperate and colder regions of the Northern Hemisphere. Most of these are found in rocky or dry soil, and some even grow in cliffs. Several species similar to the roseroot are eaten in Europe as well as in North America.

roses No flower has been more widely known and so highly praised in the civilized world as the rose (family Roseaceae, genus *Rosa*). The tally of species is difficult, if not impossible, partly because the

---

Wild Rose Syrup

Gather rose hips in late summer or autumn. Snip off both ends, then cut the hips in half. Put them into a pot and barely cover with water. Bring to a boil, cover, and simmer until the hips are tender. Strain off and retain the liquid, then cover the hips with more water and boil for 15 minutes. Strain off and retain the liquid and add it to the first batch. Measure the liquid, then add half as much sugar. Bring to a boil and simmer until the syrup thickens. Pour into sterilized bottles. Serve over pancakes.

---

rose hybridizes so easily. At least 35 species of wild rose are indigenous to North America, and of course even more wild and cultivated roses exist around the temperate regions of the world. Some wild roses even grow above the Artic Circle and in the mountains of the tropics.

All roses are edible. The hips (fruits or seedpods) are highly nutritious and can be eaten raw, in jelly, in syrup, in pies, and in teas. The flavor of the hip varies considerably, but in general is like that of apples or crab apples. As a rule, the hips of the dark red varieties of roses tend to be strong in flavor, and those of the lighter varieties are more delicate. Rose hips are extremely rich in vitamin C. When produce shipping was interrupted during World War II, the British Ministry of Health harvested 5 million pounds of rose hips from the hedges and distributed them to the populace. By weight, rose hips contain at least 20 times more vitamin C than oranges do. They are also higher in calcium, phosphorus, and iron.

Clearly, rose hips can be a good supplement to one's diet. They are especially valuable as survival food because they are available throughout most of the year. Hips develop from the flower after the petals drop off, and tend to stay on the plant through autumn and winter. Rose hips can often be found even when snow covers up most other wild foods.

Of course, the rose has also been used as a remedy for various ills. In *Historia Naturalis*, Pliny listed a dozen medicinal uses for the rose. The Apache Indians even used a wild rose tea as a cure for gonorrhea. Rose water is still available at some apothecaries and health food shops, and quite a lot of it is bottled in France. In the Middle East, rose water is sometimes kept on the tables in bottles, to be sprinkled over foods and salads. In Iran, rose water is sometimes sprinkled on the hands of arriving guests. In the Scandinavian countries, dried and powdered rose hips are marketed; the powder is sprinkled over cereals and soups, and is used for making tea. Rose hip tablets are also available in some health food outlets. In gourmet food stores, one may even find candied rose petals for sale.

Although the hips are more nutritious, the flower petals are used in candies and jellies as well as for making tea. The young leaves are also dried and used in tea, and the roots can be used for making a tealike brew as well.

S

sagebrush (*Artemisia tridentata*) Several kinds of sagebrush are native to the American West, and this rather large species grows up to 10 feet tall. It is also called big sagebrush, basin sagebrush, and wormwood. It is kin to European absinthe and tarragon, and the American Indians made use of it in various medicines. They also ground the seeds to make flour for use in making breads and chewed the leaves to help relieve stomach gas. The Indians also used sagebrush to help make an aromatic hair tonic. The pioneers used sagebrush mostly as firewood. It is quite abundant, growing from British Columbia to Mexico.

saguaros (*Cereus giganta*) Often called giant or monument cacti, these large treelike plants grow up to 50 feet in height and are the largest cacti in the United States. The central fluted stem of the saguaro may be up to 2 feet in diameter, and often one or more branches grow out at right angles and then curve upward. (Everyone who has seen many American cowboy movies knows the shape of this cactus, and has probably seen the same specimen more than once.) The saguaro has a limited range, growing wild only in southern California, Arizona, and part of Mexico. Today, however, it is somewhat rare in the wild. The saguaro has white blooms and it is the state flower of Arizona.

The saguaro bears a large crimson fruit that is delicious. The American Indians ate it fresh, made preserves of it, and even dried it for later used in a gruel. A sweet fermented syrup was made from the juice.

(See also CACTI.)

salsifies (*Tragopogon porrifolius*) Also called oyster plant, vegetable oyster, and white salsify, this species grows in the Mediterranean region in both wild and cultivated forms. (The plant is also known as goat's beard because the mature seeds, fluffy and white, resemble the beard of a goat. A similar edible plant, *Tragopogon pratensis*, is also called goat's beard.) Salsify is fairly important in French cookery, and was eaten by Louis XIV; it is marketed in Italy, Britain, and other parts of Europe. These days, most of it is raised in Belgium.

The salsify plant was introduced during colonial times to the United States, where it now grows wild and in a few gardens from coast to coast except in the southernmost states. Thomas Jefferson grew the plant for food but, in general, it has never been popular as a commercial crop in America. It is, however, becoming more widely available in markets and is available in the wild.

The whitish taproot is what is usually marketed, and it is scraped like a carrot and boiled. The young leaves can also be eaten in green salads or cooked like spinach.

Goat's beard, *Tragopogon pratensis*, grows in the same area of North America and can be eaten in the same way as salsify. In Europe, the plant *Scorzonera hispanica* is called black salsify and black oyster plant. It also has an edible taproot, which has a black skin instead of white. The roots are gathered and prepared in the same way as those of salsify.

(See also ROOT AND TUBER VEGETABLES.)

sassafras (*Sassafras albidum*) The sassafras, a medium-sized North American tree, grows from Canada to Florida, east to Missouri and Texas. It yields a rather famous tea, or spring tonic, which is made from the roots or root bark. Some years ago, sassafras tea was highly regarded even in Europe. The roots were at one time sold in specialty or health food shops, but they were taken off the market because the U.S. Food and Drug Administration deemed them to be unsafe. Apparently, the safrole in the oil is slightly carcinogenic, but most experts on wild foods point out that exorbitant amounts of sassfras would have to be consumed before it would become harmful.

In any case, the plant is still used for making sassafras tea, especially in the Appalachians. It's tasty and has a wonderful aroma. In fact, safrole was once widely used to flavor root beer, chewing gum, and toothpaste. The tree has also been called the ague tree, cinnamonwood, and smelling-stick.

The young leaves can be eaten in a green salad. Also, the dried and pulverized leaves were widely used by the Indians to flavor and thicken soups and stews. The Cajuns of Louisiana still use the leaves in the same way, and the substance is sold in supermarket spice racks under the name filé. The dried leaves, like okra, contain a mucilaginous substance that is characteristic of a true gumbo.

In addition, sassafras roots can also be used to make jelly. At one time, the plant was widely used in various folk medicines not only in America but also in Spain and England, where it was believed to cure a variety of ills, including venereal diseases. Sassafras may have been the first plant product to be exported from New England on a large scale. At one time,

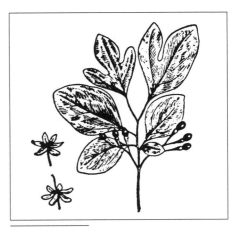

Sassafras leaves

gathering and exporting sassafras was big business. Even today some rural people take sassafras tea each year as a "spring tonic." It is believed that it thins the blood, making one ready for the heat of summer. Children still chew the stems, which have a spicy taste.

A member of the laurel family, sassafras is easily identified by its aroma and by the fact that its leaves have three or four different configurations, all on the same tree. Some of the leaves are oval, and others have two lobes. Still other leaves have one lobe, shaped like a left or right mitten.

seablite (genus *Suaeda*) Several kinds of seablite grow along salt marshes by the seas and inland salt lakes of various parts of the world. According to Michael A. Weiner's *Earth Medicine-Earth Food*, the seablite was eaten by the American Indians of the southeastern and western United States. The seeds were ground into a flour and used in breadstuffs, and the leaves were also eaten. The leaves are quite salty, however, and should be boiled in two or more changes of water.

sea grapes (*Coccoloba uvifera*) These small trees (or large bushes) grow from 25 to 60 feet high along the seacoasts of Florida, the Bahamas, and the West Indies. Often growing in dense thickets, the trees bear clusters of rounded purple fruits, which resemble large grapes. The fruits are edible, but have a large pit and are usually used to make jelly. The sea grapes' large, leathery leaves—up to 10 inches in diameter—make the plant easy to identify. The early Spanish settlers dried the leaves and used them for writing paper.

The astringent roots and bark have been used for medicine and remedies for diarrhea.

A similar fruit, the pigeon plum, *Coccoloba laurifolia*, is also edible but is not as good as the sea grape. The fruits are red and acid. This tree (sometimes called stavewood) grows up to 70 feet high.

sea kale (*Cramble maritima*) Sea kale grows wild along the shores of the Baltic and Black seas, and the western coast of Europe. Its succulent young shoots are edible, and are usually boiled in salty water or cooked like asparagus. Sea kale has a nutty flavor. The plant is sometimes cultivated in gardens. Once it is well established, a single plant can produce food for 8 to 10 years if it is properly cultivated. In addition to its use as a food, sea kale is sometimes grown as a border plant.

sego lilies (*Calochortus nuttallii*) The sego lilly, a member of the mariposa lily family, grows in the American West from South Dakota to California. Its slender stem bears a beautiful tuliplike white flower. It is, in fact, the state flower of Utah, but this honor may have come from the edible tuber instead of the flower itself. The early Mormons who settled in the Salt Lake valley ate the tubers to such an extent that the flower took on symbolic value in the Mormon Church of Utah. Reportedly, the tubers of the sego lily actually saved the lives of the Mormons during the great locust invasion of 1848.

The American Indians also ate the tubers, either by roasting them in the fire or by pounding the dried bulbs into a flour for use in mush. The tubers have a sweet taste and can be stored for long periods of time.

(See also ROOT AND TUBER VEGETABLES.)

shepherd's purse (*Capsella bursa-pastoris*) Shepherd's purse, a member of the mustard family, is also called lamb's lettuce, caseweed, pickpocket, mother's-heart, and shovelweed. It is indigenous to the Old World but has been introduced into America, where it now grows wild. The plant grows a rosette of leaves, almost flat on the ground, and, when young, these can be eaten as a potherb. Reportedly, these leaves were at one time sold at market in Philadelphia. The shepherd's purse has lots of tiny seeds—up to 40,000 on a single plant—and these can be substituted for mustard seeds. In Europe, the plant was in the past considered to have medicinal value because it helped control hemorrhoids and internal bleeding of organs. A tea made from the leaves was believed to help remedy sluggish kidneys.

Actually, there are several very similar species of shepherd's purse, and these are also edible. Some have lobed leaves rather like dandelions; others have pinnate leaves. All of the leaves are edible, but are best when young and tender.

Each plant has a stalk that grows up from the center of the rosette of leaves and develops seedpods. These seedpods are shaped somewhat like a spade, which explains the name shovelweed. They also open up to release the seeds, and thus resemble the leather pouches that the shepherds of old used to carry their food; hence, the name shepherd's purse.

silverweeds (*Potentilla anserina*) Members of the rose family, these plants grow in the northern and arctic regions. Native to Eurasia, they are now found across North America from Greenland to Alaska. In the mountains, they grow as far south as New Mexico. Called gooseworts, the plants do best along streams in salt meadows, and in damp places. They are ground-hugging plants with silvery leaves. In summer they have yellow flowers, which close at night. Silverweeds have an edible starchy rootstock that is reported to have saved entire regional populations from starvation during hard times. The rootstock can be eaten raw, boiled, or roasted, and is said to taste like parsnips.

The silverweed has also been used for medicinal purposes. It contains much tannin, which makes it effective as a gargle to treat sore throat. It was at one time used to ease menstrual cramps, and it is in fact sometimes called crampweed.

smuts These parasitic fungi attack corn, wheat, and other grains and thereby cost farmers millions of dollars each year. The common corn smut, genus *Ustilago*, causes an abnormal growth of existing cells, resulting in a large gall or tumor-like mass, which is usually black and abhorrent in appearance. Nevertheless, the American Indians (and no doubt some of the early settlers) ate the smut, usually after boiling it in water.

snowberries (*Gaultheria hispidula*) Often called creeping snowberries, these small herbs have limbs and leaves that tend to hug the ground. The plants' bright green leaves are about the size of a lady's fingernail, and their edible snow white berries are about the same size. Northern plants, snowberries grow from New Brunswick to New England, and westward along the Canadian border to the Pacific Northwest. The leaves can be eaten raw—usually as an emergency food—boiled as a vegetable or used to brew a tea. The white berries ripen in summer. They are quite good when eaten out of hand, perhaps as a trail nibble, or when used in pies, baked goods, and preserves or jelly.

The snowberry is easy to identify, and its leaves can be eaten throughout the year. It grows in forests as well as in the mountains.

Solomon's seal (family Liliaceae, genus *Polygonatum*) About 30 species of Solomon's seal grow in North America, Europe, and northern Asia. The plants have a large rootstock that has scars or markings similar to the Solomon's seal star figure or symbol. The rootstocks were used by the American Indians to make bread. Some foragers also eat the young shoots, which taste like asparagus. In Europe, Solomon's seal has been an important source of food during times of famine.

A closely related plant, the false Solomon's seal (genus *Smilacina*), grows across southern Canada and the northern United States from the Atlantic to the Pacific. This plant bears a pale red berry that was eaten by the Indians of Oregon and British Columbia. The early settlers called it the scurvy berry, and it is indeed high in vitamin C. The rootstocks of the false Solomon's seal were also eaten, usually after boiling in two changes of water. The young shoots are also edible. The Nevada Indians drank a tea made from the leaves of false Solomon's seal as a contraceptive, and the Blackfoot Indians treated wounds with powdered root of the plant.

sow thistles (*Sonchus arvensis*) These weeds, native to Eurasia, have become naturalized in America, growing over much of the United States and Canada, as well as in other places. Once established, they are difficult to get rid of because of their many seeds and their creeping root system. The leaves resemble those of dandelions, except that the sow thistle has prickly leaves and some related species of *Sonchus* even have leaf spines that must be removed before they can safely be eaten. Anyone who forages for sow thistle should take along a pair of heavy gloves for handling the leaves.

The very young leaves of sow thistle can be eaten in a salad or boiled as potherbs. As the leaves become older, their taste becomes more and more bitter. Often even the rather young plants are boiled in two changes of water. The knowledgeable forager sometimes mixes the leaves of sow thistle with other greens of more bland flavor.

Reportedly, the sow thistle has been used rather extensively in Europe as a potherb. *Larousse Gastronomique* says that the sow thistle's roots are eaten in winter. Moreover, Waverley Root claims that in Lapland the bitter milky stems are peeled and eaten raw. But these two sources give

no scientific name for the plants, and the reference could be to some other kind of thistle, perhaps of the *Cirsium* genus. The bull thistle, for example, does have edible roots and the stalks contain edible pith.

(See also BULL THISTLES; ROOT AND TUBER VEGETABLES.)

spices See HERBS AND SPICES.

spring beauties (*Claytonia virginica*) Sometimes in spring these flowers bloom in great numbers along streams and in wet places from Maine to northern Florida, west to Texas and South Dakota. Several other edible species of spring beauties grow in the same area, and still other species, such as the alpine spring beauty, *Claytonia megarhiza*, grow in the American Northwest and along the Rocky Mountains to New Mexico.

Spring beauties have edible potatolike roots, which are sometimes called fairy spuds. The roots, which grow up to almost 2 inches in diameter, are good when either boiled or baked. Fairy spuds can also be sliced and fried successfully. They are usually peeled after being cooked. The young leaves are also good either in a salad or cooked as potherbs.

(See also ROOT AND TUBER VEGETABLES.)

spruces (family Pinaceae, genus *Picea*) These evergreen trees of the pine family comprise about 40 species. They grow in all temperate and cold regions of the Northern Hemisphere. Although they are not generally eaten, they can serve as an important survival food, partly because of their availability, size, and range. Spruces grow all across Canada, for example, from the Atlantic to the Pacific.

The American Indians ate the bark of the spruce, and sometimes dried it and mashed it into flour. The young shoots were also eaten, and were a good source of vitamin C. Reportedly, some Indians sucked on the young cones as a treat, and the Crees ate the young cones as a cure for sore throat. Spruce tea, made by boiling the green needles in water, was used by the Indians and early explorers to cure or prevent scurvy. It is believed to have saved the lives of some of the men with French explorer Jacques Cartier, whose ship became icebound on the Saint Lawrence River. A hot spruce tea was also made with the twigs and cones of spruce boiled in maple syrup. This drink was used by Captain Cook on one of his voyages to help prevent scurvy among his crew. A spruce beer, sometimes brewed with the aid of maple syrup, was popular at one time and was made commercially.

The spruce is a valuable source of pulpwood and is used when a light, strong, and rather elastic wood is required. Howard Hughes used the wood to build his famous huge wood-framed airplane toward the end of World War II; the plane was dubbed the Spruce Goose.

squashes See GOURDS AND SQUASHES.

staghorn sumacs (*Rhus typhina*) In summer and autumn, these plants put forth red berries in clusters that suggest staghorns. The berries contain a large seed, and are covered with minute hairs. When washed in water, these hairs release malic acid which turns the water red and gives it the tart flavor of lemon. Drinks made from the berries are sometimes called Indian lemonade, partly because the American Indians and the early settlers made extensive use of it. The juice is also used in making jelly, sometimes being mixed half and half with the juice of elderberries.

Several other speces of *Rhus* are used for making hot or cold drinks, including the squawbush (*Rhus trilobata*), which is also used to make a syrup. In Turkey and the Caucasus a popular spice called *sumakh* is made from crushed sumac berries.

Staghorn sumac, smooth sumac, and similar species of *Rhus* are different plants from poison sumac, *Toxicodendron vernix*, which is highly poisonous and has whitish berries.

stalk vegetables Celery is the obvious example of a vegetable that is raised or gathered for its edible stalk (unless it happens to be the variety called celeriac, which is cultivated for its edible root). Other stalk vegetables include cardoon, rhubarb, and fennel. Some types of vegetables have both edible stems and stalks, such as bon choy, Chinese cabbage, and Swiss chard. Another group, such as asparagus, cattail, and bamboo, have stalks or shoots that are edible only when they are very young. Still others have edible parts or inner cores, such as burdock and young palm trees. Other stalk plants are quite fibrous but have edible juice, such as sugarcane and sorgos.

(See also ALEXANDERS.)

stone fruits The apricot, cherry, nectarine, peach, plum, and almond all belong to genus *Prunus* and have large pits or stones in the center. As a rule, the stone is discarded but in the case of the almond, it is eaten like a nut. Apricots originated in China, and still grow wild in the hills of that country. The nectarine is, of course, a smooth-skinned variety of peach, *Prunus persica nectarina*.

(See also NUTS; PLUMS; WILD CHERRIES.)

storksbills (*Erodium cicutarium*) These small plants of the geranium family grow an elongated tapering fruit that suggests a stork's bill, hence their name. The storksbill is also called alfilaria, pin clover, cransbill, heronsbill, and hemlock storksbill. The hemlock part of the latter name comes from the fact that, when young, the plant's leaves resemble those of poison hemlock. Nevertheless, the young leaves of the storksbill are edible and are sometimes available in early spring, when other greens are scarce. The leaves can be chopped and eaten in a green salad, or boiled in salted water.

The storksbill was introduced into America from Europe, and it grows wild over most of the United States. It is quite plentiful in some areas.

sunflowers About 60 edible species of sunflowers (genus *Helianthus*) grow in North America, with a few also found in Peru and Chile. The common sunflower, *Helianthus annuus*, is indigenous to the great plains of the United States. The American Indians made good use sunflower seeds, which were eaten both raw and roasted. They also obtained an oil from the seeds. The Indians also pounded the dried seeds and used them to make breadstuffs, or mixed the powdered seeds with bone marrow to form a paste or spread. In addition, they also made a spread by mixing the seeds with syrup. The shells from the seeds can be used as a coffee substitute. The early settlers used the leaves of sunflower as fodder, and the young flower heads can be boiled in salted water and eaten like artichokes or brussels sprouts. One species, the Jerusalem artichoke, is more important for its large tubers than for its seeds.

The sunflower has been cultivated for a long time, and varieties with very large heads have been developed. The seeds themselves have become popular, and they yield a good commercial cooking oil. In addition to the United States, Egypt, Russia, China, and India raise sunflowers for the oil. Also, the flowers are used for yellow dye, and the pressed oil cake (hulls from the seeds) is sold as poultry and livestock feed.

(See also JERUSALEM ARTICHOKES.)

sweet ferns (*Comptonia peregrina*) These members of the wax myrtle family—the only species in their genus—grow in rocky places in eastern North America, reaching down the Appalachians to north Georgia. The leaves of sweet ferns are used to brew a tea, and the plants also have edible nuts, which when immature can be hulled and eaten out of hand as a trail nibble or as survival food. The nuts are enclosed in a green burr.

sweet flags (*Acorus calamus*) These marsh plants grow around the world in the North Temperate Zone. Called sweet sedge, calamus, flagroot, myrtle sedge, wild iris, sweetroot, sweet rush, sweet grass, and sweet cane, sweet flag produces a rhizome that was once widely eaten as candy. Usually, the rhizome was simmered for several days in several changes of water. This boiling was necessary to tenderize the rootstock and to help remove some of the pungent flavor. When tender, the rootstock was cut into pieces and simmered in a syrup made with two parts water and one part sugar. After draining, the pieces were rolled in sugar. In addition to its use as candy, sweet flag also provides a tasty ingredient

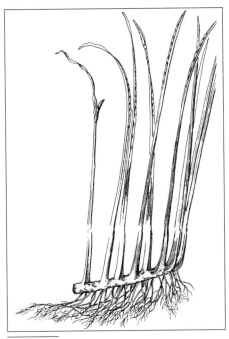

Sweet flag

for a green salad. The young stems, less than a foot high, can be peeled, revealing an inner roll of unfurled leaves, which can be eaten raw.

Be warned, however, that in recent years the U. S. Food and Drug Administration has called sweet flag unsafe to eat. In any case, it should not be confused, for eating purposes, with blue flag, which is definitely poisonous. A dependable field guide will help identify the sweet flag; the characteristic aromatic smell is a good key.

The American Indians had a number of medicinal uses for the plant—from toothache to menstrual problems—and its root has been used by the white man to cure various ills, such as stomach cramps. At one time, the powdered roots were used to help one break the smoking habit; being spicy and aromatic, the roots were supposed to help destroy the taste

for tobacco. In the past the peoples of Europe and North America used the leaves, also aromatic, as an air freshener, spreading them over the floors of churches and homes.

sweet potatoes (*Ipomoea batatas*) There is much confusion, especially in the United States, about the difference between a sweet potato and a yam. The sweet potato, native to Central America, can be cultivated as far north as New Jersey and as far south as the Andes. Some food writers call the sweet potato a yam; possibly in an attempt to clear up the confusion, at least one well-known modern cookbook writer calls *Ipomoea batatas* a boniato. In any case, the tuber is eaten mostly in the southeastern part of the United States, where it is usually called a sweet potato. As a rule, only the tuber is eaten, but it does grow edible tops, which can be cooked as potherbs. Dr. George Washington Carver promoted the sweet potato for various uses, including the tops as greens and the powdered tubers as flour for breadstuffs. Not much ever came of his work, however, and most people today use only the tubers.

The sweet potato is also raised for food in Russia and other places. It is widely eaten in Africa, along with the yam and the root of cassava. In Japan, the sweet potato has been raised on a large scale and used not only for food but also in the manufacture of starch and alcohol.

Christopher Columbus partook of sweet potatoes in the West Indies, and reported to Europe that they were grown in several varieties and were used to make a bread called *aje*, which was close to one of the local names used for the tubers. The sweet potato was also called *ages*. In the Yucatán, it was known as *camote*; the Arawak Indians called it *batatas*; in Peru, the name was *apichu*; and some Europeans called it *igname*, which also meant yams. Thus the confusion of names is nothing new.

(See also ROOT AND TUBER VEGETABLES; YAMS.)

sweetsops (*Annona squamosa*) These shrubs or small trees of tropical America and the West Indies bear edible yellowish green fruits that can reach 5 pounds. The fruits, which have a creamy pulp that is highly prized in the West Indies, are used in making soft drinks and sherbets. The pulp is sometimes frozen for the market. Sometimes called sugar apple (although it tastes more like a tart strawberry), the fruit has been introduced in the East and is raised especially in India.

A similar fruit, the custard apple (*Annona reticulata*), which has a custardlike pulp, is often confused with sweetsop and vice versa. Reportedly, this fruit was served to Columbus upon his first landing in the West Indies. Also called bullock's heart, its flesh is reddish yellow, sweet, and soft.

Also, the cherimoya (*Annona cherimola*) is frequently called custard apple. A native of the South American highlands, it is now grown commercially in Spain, Australia, New Zealand, and California. It is, however, difficult to cultivate and ship and therefore may or may not become an important fruit in the United States, although it is available on a limited basis. In any case, the cherimoya was being cultivated in the highlands of South and Central America long before the white man came.

Still another species, *Annona glabra*—known as the alligator apple, pond apple, or corkwood—also grows in South America and the West Indies. Its light wood is used like cork. Its fruits are not eaten fresh but are used for making jelly.

In spite of its name, the soursop (*Annona muricata*) is also eaten in tropical America. Its juicy pulp tastes of pineapple and mango. Like sweetsop, it is used in beverages, sherbets, and so on. It is now cultivated on its native grounds as well as in Africa, Laos, Polynesia, and India.

Finally, the atemoya is a hybrid fruit, a cross between the cherimoya and the sweetsop. It is raised commercially in Florida. It's a grayish green fruit that resembles an artichoke in appearance and is said to taste like vanilla custard with mango. According to Elizabeth Schneider's *Uncommon Fruits & Vegetables*, the atemoya is relatively high in calories and is an excellent source of vitamins C and K, potassium, and fiber.

(See also TROPICAL FRUITS.)

T

taros (*Colocasia esculenta* and related species) These plants are very important as a staple food to about half the peoples of the world—and are almost completely unknown to the other half. The plants are probably native to Southeast Asia, but have spread to most of the Pacific islands. In Hawaii and other parts of Polynesia, the famous poi paste is made from taro tubers. The taro is also eaten in the West Indies, where it is called *coco, eddo,* and *baddo.* The edible leaves are called *callaloo,* an important ingredient in a soup. The trouble is that the edible leaves of a similar plant (the malanga) are also called callaloo. One book on Caribbean cooking says that callaloo soup is also made with Chinese spinach or Indian kale.

Outside the Caribbean, the picture is even more clouded, as there are several similar plants, or varieties, eaten here and there. *Dasheen* is another popular name for taro, along with *tannia, sato imo, woo tau, hung nga woo tau,* and others. In any case, the taro is not a yam and not a cassava, although it is similar in some ways to both of these tropical edible roots. It is a member of the arum family.

Both the leaves and shoots of the taro are edible, but the root is the important part. It is eaten in soups and stews or fried in fritters, and is even eaten in desserts. The plant (in one variety or another) is now cultivated in the Caribbean, South America, Central America, Africa, the Pacific islands, the East Indies, and probably elsewhere in tropical climates. More and more it is becoming available in the United States, especially in ethnic neighborhoods.

As mentioned above, a very similar plant, malanga (also called yautia, cocoyam, etc.) comes in 40 species of genus *Xanthosoma.* All are native to tropical America. These plants are eaten mostly in Cuba and Puerto Rico. Adding to the confusion, taro is called *malanga isleña* in Cuba. When fresh, malanga has crisp flesh, similar to water chestnuts. The tubers are usually boiled, or used in fritters, pancakes, and stews.

(See also ROOT AND TUBER VEGETABLES.)

teff (*Eragrostis abyssinica*) An important Abyssinian grass with some ornamental value, teff produces edible seeds, which are usually ground into flour. Teff was an important grain in Ethiopia, where it is still eaten. The grass is also cultivated and marketed as a grain (or flour) on a limited basis in other countries, including the United States. The flour is said to be high in iron, protein, calcium and zinc.

(See also GRAINS AND GRASSES.)

ti plant (*Cordyline terminalis*) Also called turf trees and cabbage trees, these tropical plants are cultivated in parts of Asia, and are especially plentiful on some of the Pacific islands, where they are found both in the wild and cultivated. The plants, which grow up to 15 feet high, have large, leathery leaves that are often used for wrapping food that is to be cooked in heated pits in the ground. Ti plants have an edible rootstock that is high in starch and that is usually baked. The roots are also

Ti plant with enlarged view of leaves.

used for making soft drinks. Furthermore, ti plants also bear red berries that are edible when ripe.

tomatoes (*Lycopersicon esculentum*) Native to South America, tomatoes were under cultivation as far north as Mexico when the Spanish explorers arrived. The Spanish took the tomato to Europe, but it was believed to be poisonous by some botanists of the day. (The leaves and stems can, in fact, be toxic.) As late as the 19th century, the tomato was accused in England and the United States of causing cancer. But in time it became accepted, and today it is widely eaten in Europe, America, and elsewhere. The

Fried Green Tomatoes

This dish, long enjoyed in the southern part of the United States, has recently been in the limelight as the title of a popular movie and book. Usually, only the red tomatoes (or those in the process of turning red) are marketed in supermarkets, but this may change. In any case, fried green tomatoes are a wonderful way for home gardeners to start enjoying tomatoes before they are ripe. To proceed, merely pull a few mature but still green tomatoes off the vine and cook as follows:

4 green tomatoes, sliced crossways 1/4-inch (0.64-centimeter) thick
cooking oil
1/2 cup (113 grams) flour or fine cornmeal
1 teaspoon salt
1/4 teaspoon black pepper

Mix the flour or cornmeal, salt, and pepper. Dust the slices of green tomato. Heat some cooking oil in a skillet. Fry tomato slices on medium heat, turning once, until browned on each side. Drain on a brown grocery bag and serve.

There are other recipes, some calling for chicken egg and other ingredients. The one above has been adopted from *The Progressive Farmer's Southern Cookbook*, published some years ago. Finely ground meal from whole corn is required, but not all supermarkets carry the right stuff (which is usually labelled water ground or stone ground), in which case ordinary flour is the best choice. *The Progressive Farmer* book recommends using bacon drippings as cooking oil, which of course is a good way to use up the drippings left from breakfast bacon. But be warned that modern health food experts recommend vegetable oil over animal fat for frying foods—and some recommend total abstinence from fried fare.

Italians have made exceptional use of the tomato in their cuisine.

The fruit is eaten raw or cooked when ripe, and is also cooked when green, as in fried green tomatoes and other recipes. Tomatoes don't freeze very well but they are widely canned. Of course, they are used in juice and catsup as well as in chili sauce and barbecue sauce. They are increasingly being marketed in dried form.

There are a number of varieties in cultivation, such as the small cherry tomato, the Mexican green husk tomato (*Psysalis ixocarpa*) and the tree tomato (*Cyphomandra betacea*); these are different species.

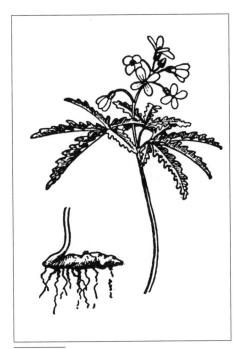

Toothwort

toothworts (*Dentaria laciniata*) Sometimes called cutleaf toothworts, pepperroots, or crinkleroots, these plants—growing up to 15 inches high—are plentiful in moist woods from Quebec to north Florida and generally west to the Mississippi River. The plants bear an edible rootstock that can be eaten raw in salads, or grated and used like horseradish. In *Free for the Eating*, Bradford Angier recommends a sauce made from grated toothwort root mixed with a little vinegar and served with boiled moose. Angier also lists the two-leaved toothwort, *Dentaria diphylla*, as being equally tasty. Other species are also edible.

(See also ROOT AND TUBER VEGETABLES.)

trees See ACACIAS; ASPENS; BARK; BAOBABS; BEECH TREES; BIRCH TREES; BLACK LOCUSTS; BUTTERNUTS; CALIFORNIA LAURELS; CAROBS; CHESTNUTS; COCOA; COCONUT PALMS; DATE PALMS; DURIANS; ELDERS; FEIJOAS; HACKBERRIES; HAZELNUTS; HEMLOCKS; HICKORIES AND PECANS; JUDAS TREES; JUNIPERS; LARCH TREES; MAPLES; MESQUITES; MOUNTAIN ASHES; NETTLE TREES; PALMS; PINES; PINONS; SASSAFRAS; SPRUCES; WALNUTS.

tropical fruits Bananas ship well and are available all year, which make them seem ubiquitous in the supermarkets of the world. The pineapple is also widely available, either fresh or canned, or as juice. (The pineapple, by the way, is usually associated with Hawaii, but is native to Central and South America and the Caribbean.) The kiwi also has a long shelf life and has become increasingly popular in recent years. The mango is highly important to some people who live in the tropics, where it is a part of their diet and cuisine. The mango has been called the apple of the tropics, but it does not ship well and is not widely available in other parts of the world, although it can be found in some markets.

A number of more or less edible fruits are called *zapote* or *sapote* in Middle America, but most of these are not widely available to the rest of the world; the white sapote (*Casimiroa edulis*) is being raised on a limited basis in Florida and California, but a market has not yet developed, partly because the fruit is of inconstant quality. The species grow quite large, however, and a single tree may produce up to 5,000 pounds of fruit per year. A number of fruits of family Sapotaceae are eaten on the various islands of the Caribbean and in Central America, but again, it's hard to tell exactly which is which. Part of the problem is that the Mexican word *zapote* comes from an Aztec word *tzapotl*, meaning simply any sort of soft fruit; the other part of the problem is that people who market the various tropical fruits call them whatever name they think will sell the product.

truffles Edible subterranean fungi, truffles are found mostly in Europe, and are especially sought in France and Italy. They normally grow wild, attached to the roots of trees. The oak is usually the most productive kind of tree, but in England one species of truffle grows mostly on beech roots.

Usually, the truffle is about a foot under the ground and is difficult to detect. Pigs and dogs are fond of rooting truffles and can find them by smell; some of these animals are actually trained to hunt them. Some keen-nosed people have been able to smell out truffles, but this is not common. Skilled hunters, will, of course, have secret spots, and will know that truffles do best in dirt that

contains quite a few rocks, which will provide good drainage. Some truffles grow as big as oranges, and the larger ones sometimes cause cracks in the ground, which aids in detection.

Several species of truffle grow in France. One of the more highly prized is the black Périgord truffle (*Tuber melanosporum*), found near Périgord and elsewhere in the country. The truffle has economic significance in France, and an attempt has been made by the government—with some success—to reforest some barren areas with truffle-bearing trees. It is possible to "cultivate" truffles on new ground by planting oak seedlings from trees that have the fungus. After such a seedling is planted, it may take several years for full production.

The forests of northwest Italy grow a famous white truffle, called *tartufo bianco*, that is highly prized. Often the white truffle is sliced and eaten raw. A black Italian truffle, grown in the same area, is called *tartufo nero*. These were highly prized by the ancient Romans. (Pliny held the truffle to be among the most wondrous of all things.) They are cooked in omelets and chopped in spaghetti. In England, the principal truffle is *Tuber aestivum*, which is found mostly in beech woods. Other edible truffles grow here and there around the world, including the desert truffle in North Africa. A few truffles also grow wild in America, mostly in Oregon and California, but they are scarce.

tuber vegetables See ROOT AND TUBER VEGETABLES.

tulips (family Liliaceae, genus *Tulipa*) Tulips are believed to have originated in ancient Persia, and one species still grows wild in Asia Minor and across north and central Asia—all the way to Japan. A number of species grow wild, and, of course, many variations have been developed in the flower industry.

The petals as well as the stems of the tulip are edible. Generally, the flower is too expensive to be used as food, but it was eaten in Holland during World War II.

V

violets (*Viola canadensis* and other species) These beautiful flowers grow wild in many parts of the world. Often they are the first plants of spring, which makes them a valuable forage species. The young leaves can be nibbled raw, put into a salad, or cooked as potherbs. The ancient Egyptians grew violets in pots and used the leaves in salads and as seasoning. In France and Italy, the shoots are put in salads, and some modern practitioners use the flower petals in salads for color as well as flavor. The Greeks used violets as a medicine to ease stress and promote sleep. And, believe it or not, the ladies of Old England made a face cream of violets and goats' milk.

A tea is often made with the dried leaves, and the blossoms can be used to make jelly and syrup. In times past, all manner of curative powers were attributed to one sort of violet or another. The ancient Greeks, for example, believed that violets could cure insomnia.

In recent years, it has been suggested that violets should be eaten in moderation, and that some may have a cathartic effect.

W

walnuts (genus *Juglans*) Some 17 species of walnuts grow in various parts of the world. The best known is the Persian or English walnut, *Juglans regia*, which has been so widely cultivated that nobody knows its exact origin. The British imported it from Persia (perhaps with the aid of the Romans) and called it the Persian walnut. The Americans got it from England, and called it the English walnut. In any case, this species has a thin shell and lots of meat, which is part of the reason that it is so successful commercially. These and other walnuts were very important food to parts of Europe during times of famine and, it has been reported, even the shells were ground with acorns to make bread. The meats were also pulverized and mixed with a little water to form a kind of milk. Today, the meats are used like other nuts in some areas, and they are widely used in the cuisine of the Middle East. In Bulgaria, for instance, walnut and cucumber soup is popular.

In North America, the native black walnut (*Juglans nigra*) grows wild and is sometimes gathered for market, especially in the Ozarks. The nuts of this species are thick, don't have much meat, and are difficult to shell. They also have a much richer flavor that makes them a little too strong for eating out of hand. The nuts are, however, highly valued in some quarters for use in candy, ice cream, and other sweets. The American Indians ground the meats into a flour, and also made a sweet syrup from the sap from the black walnut tree.

Juglans nigra grows mostly in the eastern part of North America, but several related species grow in the West and Southwest. The trees have been used for timber as well as for such items as gunstocks. During World War II airplane propellers were made from the wood of the black walnut. The North American butternut is also a species of *Juglans*. (See also BUTTERNUTS; NUTS.)

watercress (*Nasturtium officinale*) The watercress, an aquatic member of the mustard family, is native to Europe and Asia. It has now been introduced to North and South America and to the West Indies. In the United States, it now grows wild in every state, including Alaska and Hawaii, and in some places it can be gathered during every month of the year. Before it became Rocket City USA, Huntsville, Alabama, billed itself as the watercress capital of the world.

Watercress

261

The plant grows in cool streams, and in shallow water it takes root in suitable bottom.

The watercress is a favorite plant for foragers (and anglers) because it is easy to spot, gather, and prepare. It also tastes good, although it is slightly pungent. It is used raw in salads, cooked as a potherb, or stir-fried. It can also be cooked in stews and soups, chopped and scrambled with eggs, or eaten in sandwiches. Tea can be made from the leaves and roots.

Although the plant is not poisonous, it should of course be taken from clean water, especially if it is to be eaten raw. Some guidebooks recommend that the leaves be washed in water that has been treated with water purification tablets. Remember also that the highly poisonous water hemlock also grows along streams. While the two plants don't resemble each other, positive identification should be made before eating anything that looks as if it might be watercress.
(See also GREENS.)

water lettuces (*Pistia stratiotes*) An aquatic plant of the arum family, water lettuce grows in the tropical and subtropical regions of Asia, Africa, and America, including Florida. It's a floating plant that seldom takes root. As its name implies, it resembles young lettuce, but does not form a head. The plant

Water lettuce

usually grows in lakes, ponds, and backwaters, where it tends to form large patches. It seldom thrives in a stream.

Although water lettuce is quite tender and tempting to use in a tossed salad, it is best to boil the leaves before eating them. (See also GREENS.)

watermeal (*Wolffia brasiliensis*) These tiny members of the duckweed family float on the surface of the water and look like specks of green meal. Rootless, the individual plants are no longer than 1 millimeter long. Watermeal, along with other tiny duckweeds, are one of the most common and widespread of all plants. Various species are harvested to feed cattle and pigs in Africa, India, and Southeast Asia. Also, watermeal is eaten by people in Burma, Laos, and Thailand, where it is even cultivated as a vegetable called *khai-nam* (meaning "egg of the water").

water shields (*Brasenia schreberi*) These water lilies have small flowers (less than an inch across) and round leaves that float on the water. By comparison, other water lilies have large, showy flowers and larger leaves, or pads, that have a cleft instead of being round. The stem of the water shield grows to the center of the leaf.

The water shield grows in ponds and slow streams in most parts of the wild. Some of the American Indians ate the roots of the plant, and the Japanese eat the young unfurling leaves and stems as a salad. The parts of the plant under the water have a slimy surface, and the young shoots are coated with mucilage. This substance is also eaten.

wild cherries Dozens of species of cherry trees, which bear edible fruits with large, round stones in the center, grow in the temperate regions of the world. Several species grow wild in North America, as well as in Europe, and many species or varieties are cultivated for fruit and beautiful flowers. Some of the cherry trees also have value as wood for building furniture.

In America, the black cherry (*Prunus serotina*) is one of the more popular native species, growing wild in the eastern half of the United States. The fruits of this tree have been used in flavoring brandy, and they are sometimes called rum cherries. These and other wild cherries have also been used to flavor commercial cough syrup, and the bark as well as the fruits has been used in cough remedies. The fruits of wild cherries have also been used to make wines, jams, and jellies. The American Indians made good use of wild cherries, sometimes drying them for use in pemmican.

In addition to truly wild cherries, several varieties have been imported from Europe for cultivation in America, and some have escaped the grove or garden and now grow wild. In short, the quality of wild cherries varies greatly from tree to tree, and some species, such as the common chokeberry (*Prunus virginiana*), are not as tasty as others for eating as a fruit. Be warned that at least some species of wild cherries contain poisonous hydrocyanic acid in both the leaves and in the fruit pits. Nevertheless, wild cherry leaves have been dried and smoked as a substitute for tobacco.

Herodotus, the ancient Greek historian, spoke of people in a distant land, somewhere in the edges of Scythia, who lived on fruit that are believed to be wild cherries. As he described:

The tree from which they get their food is called Pontic tree: it is about the size of a fig-tree and bears a fruit as big as a bean and it has a stone inside it. When this fruit is ripe, they strain it through cloth, and the juice that comes out is thick and black, and they call it *aschy*. The juice they lick up or mix with milk and drink it, and from the thick part, or lees, they make cakes to eat; for they have but few cattle, the pasture being far from good. Every man makes his home under a tree, covered in winter with a cloth of white felt and uncovered in summer. No man harms these people, for they are regarded as holy, and they own no weapons of war. (See also STONE FRUITS.)

wild ginger (*Asarum canadense*) Sometimes called Canada snakeroot and colicroot, this small herb of the birthwort family grows in the northeastern part of North America, and a few other species grow in other regions. A similar species in Europe (*Asarum europeum*) is called asarabacca. The plant has a large aromatic rootstock that smells and tastes similar to commercial ginger. Wild ginger can be used as a spice to flavor meats and other foods or to make gingerbread, candy, and syrup.

In Europe, asarabacca was at one time cultivated for medicinal purposes, and the American Indians used wild ginger for several ailments and conditions, including earache.

The spicy dried ginger, the root of *Zingiber officinale* of tropical Asia, is used over much of the world to flavor foods, as are the fresh roots. Ginger was an important item of trade for the early New Englanders who, according to Waverley Root, ate candied ginger after a meal to "aid digeston and curb flatulence." Ginger is now raised not only in Asia but also in Australia, South Africa, Jamaica, and other lands.

(See also ROOT AND TUBER VEGETABLES.)

wild potato vines (*Ipomoea pandurata*) These members of the morning glory family grow over the eastern parts of the United States from New York to north Florida and west to Texas. They are perennial vines that thrive along roadways, fencerows, and other open or partly shaded areas. The large white flowers grow up to 3 inches wide; and the edible roots, which look rather like sweet potatoes, grow up to 20 pounds. Usually, the young roots make better eating, as the large ones are often too woody. It's best to dig the roots in summer or early autumn, but they can be taken at any time as emergency food. The roots can be sliced and dried for future use, or they can be frozen.

The young roots can be boiled in water or baked like a potato. If the roots are bitter, however, it's best to boil them in two or three changes of water. To serve, peel the cooked roots and serve them with butter, salt, and pepper.

Out west, the bush morning glory (*Ipomoea leptophylla*) grows on a bush about 4 feet tall and has a root weighing up to 25 pounds. The plant is sometimes called either manroot or bigroot. Its root is edible, but is not usually described as toothsome.

(See also ROOT AND TUBER VEGETABLES.)

wild rice (*Zizania aquatica*) Often called Indian rice, this well-known grass grows in shallow water over a rather muddy bottom. It thrives in brackish waters as well as fresh, and it may be found along the shores of streams or lakes as well as in swampy places. Its natural range is from the south of Canada to northern Florida and west to the Mississippi. It also extends over North Dakota, Montana, the panhandle of Idaho and Oregon. A separate species (*Zizania texana*) grows in Texas. It has been introduced in many areas by sewing seeds in suitable locations, and it is widely planted as a natural food to attract ducks. According to the late Euell Gibbons (author of *Stalking the Wild Asparagus*), wild rice expands faster when it is harvested. (It does best in from 3 to 4 feet of water. Seeds and instructions for planting wild rice can be obtained from catalogs that traffic in foods for wildlife.)

When properly prepared, wild rice is delicious and nutritious. The Indians gathered large quantities of it, and it has been marketed for a long time. Today, its price is too high for daily consumption by most of the populace. As sold in supermarkets, it is often mixed in with regular rice. Although it is still gathered by modern foragers for home use, it was the American Indians who made it a part of their culture in some areas. Here's a report from Michael A. Weiner's *Earth Medicine-Earth Foods*:

> The large, plume-topped grass wild rice is found growing in ponds and swamps and along the marshy borders of streams throughout the central and eastern states. The seeds were a very important cereal food among the Ojibwa and Chippewa Indians. The Menominee tribe was even named for the dependence upon this wild food. They called themselves *Menomin*, or "wild rice men," because they lived mainly on the wild rice of the lakes of their region.
>
> Indian women did the rice gathering. Before the rice was ripe, they paddled among the bunches, binding several tall rice plants together every

few feet. The women returned to the rice on the day of ripening and collected the ripened grain by bending the bound shocks over their canoes and beating them with a stick. Whatever few grains fell outside of their canoes became the next generation of rice and was gathered during the next season of collection.

The Indians prepared wild rice by parching the seeds in a receptacle for a short while over hot coals. The rice was stirred constantly to keep from burning. When cool, the husks were beaten off and the seeds winnowed. The parched seeds were then boiled in water and eaten with blueberries or maple syrup or used for thickening soups. Like the commercial variety, wild rice swells with boiling, sometimes increasing in size from three to four times during cooking.

The Indians recognized the high food value of this species and many ceremonies were devoted to the cult of the wild rice plant. The Menominees held a yearly thanksgiving festival when the great Manitou was thanked for providing this food for his people. This ceremony is held just after the rice has been gathered and before the fall hunts begin.

(See also GRAINS AND GRASSES.)

wild strawberries (genus *Fragaria*) Cultivated strawberries need no introduction here, but it should be pointed out that the Indians of Chile were cultivating strawberries of exceptional size—reported to be as big as walnuts—when the white man arrived in the New World. The cultivated strawberries of today are probably a hybrid between this large South American berry and a wild meadow strawberry of North America, both of which were taken to Europe. Several other species grow in the tem-

perate regions of the world, and some wild species grow in North America.

The taste and size of wild strawberries vary widely from one place to another and from one plant to another. In general, however, they can be eaten out of hand or used in recipes just like cultivated strawberries. Some wild berries have even more flavor than the cultivated ones. They make excellent jams, jellies, and pies. The leaves of wild strawberries can be used to brew a tea, which is high in vitamin C. According to *Foxfire 2*, the leaves of wild strawberry, along with those of blackberry, were once eaten after boiling them in water with a little fatback for seasoning.

In Europe, the wild strawberry has been used in medicines, and the famous botanist Carolus Linnaeus (who was also a physician) ate the berries to keep himself free of gout.

willows Well over 100 species of willow plants (genus *Salix*) grow in cold and temperate regions in both the Northern and the Southern hemispheres. Some are trees, but most are rather small, and in the Arctic they may grow only a few inches high. About 70 species are native to North America. Throughout their range, willows prefer low land along streams and lakes, or swamps.

Although willows are a favorite browse for deer and rabbits, they are not normally eaten by man. Nonetheless, their high nutritional value should not be overlooked. They are among the first sources of vitamin C to sprout forth in spring. The young shoots, when peeled, can be eaten raw. The young leaves can also be eaten raw, preferably in a salad with other greens. The leaves can also be boiled or steamed.

The inner bark of the willow can also be eaten, and can be dried and ground into a flour. Also, the inner bark of some species contains a bitter substance called salicin, which has medicinal functions similar to those of aspirin and can be used to reduce fevers.

(See also ASPENS.)

wintercress (*Barbarea vulgaris and Barbarea verna*) This native of Europe now grows in the northern part of the United States, south to Georgia. It also grows in Washington State and Oregon. The plant has for a long time been used as a potherb, partly because it grows during cold weather and can sometimes be gathered when no other greens are available. In *Stalking the Wild Asparagus*, Euell Gibbons wrote: "Where I live in suburban Philadelphia, the first sign of spring is not the returning wild geese winging high, nor the robins of the lawn. These harbingers are always preceded by the Italians, swarming out from town to gather wintercress from fields and ditches."

The plant is also called Herb of St. Barbara because it is often the only potherb that can be gathered fresh on St. Barbara's day, December 4th. Other names include upland cress and yellow rocket.

The leaves of wintercress are boiled as potherbs, and are tasty when eaten before warm weather sets in. The plants put forth lots of yellow flowers, whose buds can be cooked and eaten like broccoli. Be warned, however, that eating wintercress may cause a kidney malfunction.

Y

yams (genus *Dioscorea*) These plants probably originated in Africa, but no one knows for certain. Whatever their origin, there are now over 600 species in the tropical and subtropical regions of both hemispheres, and several kinds are cultivated for food. Many yams produce only one tuber, growing from the base of the stem or stalk, but some kinds do produce several tubers, as do sweet potatoes.

On the whole, the yam is much more important than the sweet potato. The yam is in fact one of the most important of all cultivated crops. Yet, the true yam is almost unknown in the United States, where the term "yam" is commonly used incorrectly to denote a sweet potato. The two tubers are similar, and some yams do at least look like a sweet potato. As a group, however, the yams grow much larger, and in some areas they even have cultural and spiritual significance apart from food value. Here's a report from Elizabeth Schneider's *Uncommon Fruits & Vegeables*:

> Not only do yams have value as sustenance for a vast number of people, they are also revered as religious objects and given ceremonial status—as was life-giving corn for some American Indians. The Trobriand Islanders built intricately decorated wooden "yam houses" where the splendid tubers are ensconced to be viewed by neighbors. In Cuba, yams are considered festive food, to be saved for special occasions. Because they can stay in the ground longer than just about any other tuber, they can be hauled out and appreciated whenever important celebrations occur—and for these, the bigger the

better. It may well be that yams are worshiped because they can become awesomely huge. In the Pacific Island of Ponape, the size of yams is described as 2-man, 4-man, or 6-man, designating the number needed to lift the tuber. One would be tempted to dismiss this as an exaggeration were it not for the fact that 600-pound yams 6 feet long have been recorded.

Yams can be cooked in a number of ways—fried, boiled, whole, and sliced. They are often grated for use in all manner of dishes, including sweet puddings. There is also evidence that in voodoo rituals yams were at one time cooked in a stew or soup with the heads of sacrificed humans and congo beans. Be that as it may, yams grow wild in many tropical parts of the world, especially in West Africa. The tubers of wild yams can be used as food, although some certainly aren't as palatable as others, and the leaves can also be eaten. Some yams also grow air potatoes.

(See also AIR POTATOES; ROOT AND TUBER VEGETABLES; SWEET POTATOES.)

yeasts This term applies to a group of tiny plants that have the ability to turn sugar into carbon dioxide and alcohol. Under the proper conditions, yeast grows very rapidly, and when added to bread dough it multiplies so fast and generates so much carbon dioxide that it causes the bread to "rise." Thus, human beings eat trillions of yeast plants in bread, and consume trillions more in beer and other fermented beverages. Although yeasts aren't considered to be "food," some connoisseurs of sourdough bread will argue that one yeast certainly is not good as another. In any case, yeasts are widely available in cakes or packets in modern supermarkets and they have been very important in the development of man's food and drink. Yeast was probably used quite early in human history, and one species (*Saccharomces winlocki*) has even been found in the sediment in a Theban beer jar dating back to 2000 B.C.

In addition, yeast has been touted as having medicinal value, and was once thought to cure the plague. Even Hippocrates, the father of modern medicine, believed that yeast had curative powers.

yuccas A genus of the family Liliaceae, yuccas comprise at least 30 species. Many have edible flowers, which grow profusely on a tall stalk. The flowers are eaten either raw or cooked. Some species bear an edible fruit. The seeds are also edible, and in spring the core of the young flower stalk can be eaten raw or cooked.

The Indians of the southwestern United States and Mexico made extensive use of yucca plants for food and other purposes. The bladelike leaves made good cordage fiber and were also used for weaving baskets, mats, and chair seats. The roots of some species were crushed and used for soap. Reportedly, the Hopi washed their hair with yucca root and then rubbed in duck grease to prevent baldness. One species of yucca is even called soapweed (*Yucca glauca*),

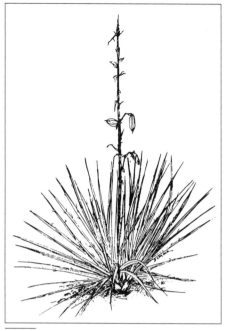

Yucca

which seems to grow in a band up the middle of the United States from the Big Bend of Texas to North Dakota.

Because of the shape of the pointed leaves, the yucca has also been called Spanish bayonet, and in fact has been planted for use in fortifications. Other names for the plant include Adam's needle, desert candle, and Quixote plant.

Z

zamias This genus is made up of about 30 species of fernlike plants that grow in the American tropics, including the *Zamia floridana* and *Zamia pumila* of south Florida. The stems are said to be a source of starch, used after the alkaloid has been washed out.

At one time, the zamia was eaten in the West Indies. According to Reay Tannahill's *Food in History*, the Spaniards found that the Taino Indians had a culinary specialty called "zamia bread," which required some doing:

It was made by grating the stems of the zamia plant, then shaping the pulp into balls and leaving them in the sun for two or three days until they began to rot and turn black and wormy. When sufficiently ripe, the balls were flattened into cakes and baked over the fire on a griddle. If the bread was eaten before it became black and wormy, said the Tainos, "the eaters will die." In this they were right; unless zamia pulp is either intensively washed or fermented, it can be toxic.

In Florida, zamia is called comfortroot and coontie. According to a booklet entitled *Seminole Indian Recipes*, bread made from coontie root was a staple of the Seminole Indian's diet.

Additional Reading

Angier, Bradford. *Free for the Eating*. Harrisburg, Pennsylvania: Stackpole Books, 1966.

Boswell, John, ed. *The U. S. Armed Forces Survival Manual*. New York: Wade Publishers, 1980.

Coon, Carleton S. *The Hunting Peoples*. New York: Nick Lyons Books, 1971.

Dobelis, Inge N., project ed. *Magic and Medicine of Plants*. Pleasantville, New York: Reader's Digest Association, 1986.

Elias, Thomas S., and Dykeman, Peter A. *Field Guide to Edible Wild Plants*. New York: Outdoor Life Books, 1982.

Gibbons, Euell. *Stalking the Wild Asparagus*. New York: David McKay Company, 1962.

Livingston, A.D. *Outdoor Life's Complete Fish & Game Cookbook*. Danbury, Connecticut: Outdoor Life Books, 1989.

Macdonald, David, ed. *The Encyclopedia of Mammals*. New York: Facts On File, 1987.

McClane, A. J., ed. *McClane's New Standard Fishing Encyclopedia*. New York: Holt, Rinehart and Winston, 1965.

Root, Waverley. *Food*. New York: Simon & Schuster, 1980.

Schneider, Elizabeth. *Uncommon Fruits & Vegetables*. New York: Harper & Row, 1986.

Smith, Leona Woodring. *The Forgotten Art of Flowery Cookery*. Gretna, Louisiana: Pelican Publishing Company, 1985.

Tannahill, Reay. *Food in History*. New York: Crown Publishers, 1988.

Weiner, Michael A. *Earth Medicine-Earth Foods*. New York: The Macmillan Company, 1972.

Index